System Requirements Specification (SyRS) 2.0

--The Structure-Behavior Coalescence Approach--

William S. Chao

Structure-Behavior Coalescence

Systems Architecture = **Systems Structure** + **Systems Behavior**

CONTENTS

CONTENTS 5

PREFACE 9

ABOUT THE AUTHOR 11

PART I: BASIC CONCEPTS 13

Chapter 1: Introduction to System Requirements Specification....15
1-1 Systems Development Life Cycle................................15
    1-1-1 Project Planning 16
    1-1-2 Requirements and Specifications 17
    1-1-3 Design and Implementation 17
    1-1-4 Verification and Validation..........................17
    1-1-5 Product Evolution...................................17
1-2 System Requirements Specification 18
1-3 Multiple Views of a System................................19
1-4 Multiple Views Non-Integrated Approaches for Systems Requirements Specification 1.0 23
1-5 Multiple Views Integrated Approaches for Systems Requirements Specification 2.0 26

Chapter 2: Systems Structure and Systems Behavior....................29
2-1 Structure of Systems 29
2-2 Behavior of Systems 35

Chapter 3: Structure-Behavior Coalescence 43
3-1 Integrated Whole to Achieve the System Requirements Specification..43
3-2 Integrating the Systems Structure and Systems Behavior 46

3-3 Structure-Behavior Coalescence to Facilitate an Integrated Whole 46

3-4 Structure-Behavior Coalescence to Achieve the System Requirements Specification 47

3-5 SBC Approach for System Requirements Specification 2.0 50

3-6 SBC Model Singularity...51

PART II: SBC APPROACH FOR SYSTEM REQUIREMENTS SPECIFICATIONS 2.0 55

Chapter 4: Architecture Hierarchy Diagram 57

4-1 Decomposition and Composition...57

4-2 Multi-Level Decomposition and Composition 59

4-3 Aggregated and Non-Aggregated Systems..............................63

Chapter 5: Component Operation Diagram 65

5-1 Operations of Each Component...65

5-2 Drawing the Component Operation Diagram 70

Chapter 6: Interaction Flow Diagram 73

6-1 Individual Behavior Represented by Interaction Flow Diagram 73

6-2 Drawing the Interaction Flow Diagram 74

PART III: CASES STUDY...81

Chapter 7: System Requirement Specification 2.0 of the Smart Parking Cloud Applications and Services IoT System 83

7-1 Architecture Hierarchy Diagram 84

7-2 Component Operation Diagram..86

7-3 Interaction Flow Diagram 101

Chapter 8: System Requirement Specification 2.0 of the Landslide Prevention and Relief Cloud Applications and Services IoT System 111

8-1 Architecture Hierarchy Diagram 112

8-2 Component Operation Diagram..114

8-3 Interaction Flow Diagram..127

APPENDIX A: SYSTEM REQUIREMENTS

SPECIFICATION 2.0 ....**139**

APPENDIX B: SBC PROCESS ALGEBRA **143**

BIBLIOGRAPHY **147**

INDEX...**153**

PREFACE

System requirements specification (SyRS) is, in the systems development life cycle, a result of the requirements and specifications phase. That is, a system requirements specification is for the analysts to find out what the customers indeed expect the system to do for them. When working on the system requirements specification, we only specify what this system is, but never ask how this system shall be manufactured.

A system has been specified, by system requirements specification (SyRS) 1.0, hopefully to be an integrated whole, embodied in its assembled components, their interactions with each other and the environment. Since systems structure and systems behavior are the two most prominent views of a system, integrating the systems structure and systems behavior apparently is the best way to achieve a truly integrated whole of a system. Because system requirements specification 1.0 does not specify the integration of systems structure and systems behavior, very likely it will never be able to actually form an integrated whole of a system.

Structure-behavior coalescence (SBC) provides an elegant way to integrate the systems structure and systems behavior, and hence achieves a truly integrated whole, of a system. A truly integrated whole sets a path to achieve the desired system requirements specification (SyRS). SBC facilitates an integrated whole. Therefore, we conclude that system requirements specification (SyRS) 2.0 using the SBC approach, which contains three fundamental diagrams: a) architecture hierarchy diagram, b) component operation diagram and c) interaction flow diagram, is highly adequate in specifying a system.

ABOUT THE AUTHOR

Dr. William S. Chao is the CEO & founder of SBC Architecture International®. SBC (Structure-Behavior Coalescence) architecture is a systems architecture which demands the integration of systems structure and systems behavior of a system. SBC architecture applies to hardware architecture, software architecture, enterprise architecture, knowledge architecture and thinking architecture. The core theme of SBC architecture is: "Architecture = Structure + Behavior."

William S. Chao received his bachelor degree (1976) in telecommunication engineering and master degree (1981) in information engineering, both from the National Chiao-Tung University, Taiwan. From 1976 till 1983, he worked as an engineer at Chung-Hwa Telecommunication Company, Taiwan.

William S. Chao received his master degree (1985) in information science and Ph.D. degree (1988) in information science, both from the University of Alabama at Birmingham, USA. From 1988 till 1991, he worked as a computer scientist at GE Research and Development Center, Schenectady, New York, USA.

PART I: BASIC CONCEPTS

Chapter 1: Introduction to System Requirements Specification

A system requirements specification (SyRS) has traditionally been viewed as a document that communicates the requirements of the customer to the technical community who will specify and build the system.

For the SyRS approach to specify a system as an integrated whole of that system's multiple views, it must be able to integrate the systems structure and systems behavior when specifying a system.

Current multiple views non-integrated approaches for system requirements specification 1.0 such as data-oriented, function-oriented, control-oriented and object-oriented, more or less, fail to specify a system as an integrated whole of that system's multiple views because they are not able to integrate the systems structure and systems behavior when specifying a system.

Multiple views integrated approaches for system requirements specification 2.0, such as SBC, provide a sophisticated way to integrate the systems structure and systems behavior when specifying a system.

1-1 Systems Development Life Cycle

The phases of the systems development life cycle (SDLC) [Blan08, Koss11, Wald15], as shown in Figure 1-1, are: a) project planning, b) requirements and specifications, c) design and implementation, d) verification and validation and e) product evolution. The systems development life cycle applies recursively to life cycles that produce hardware and software portions of the system.

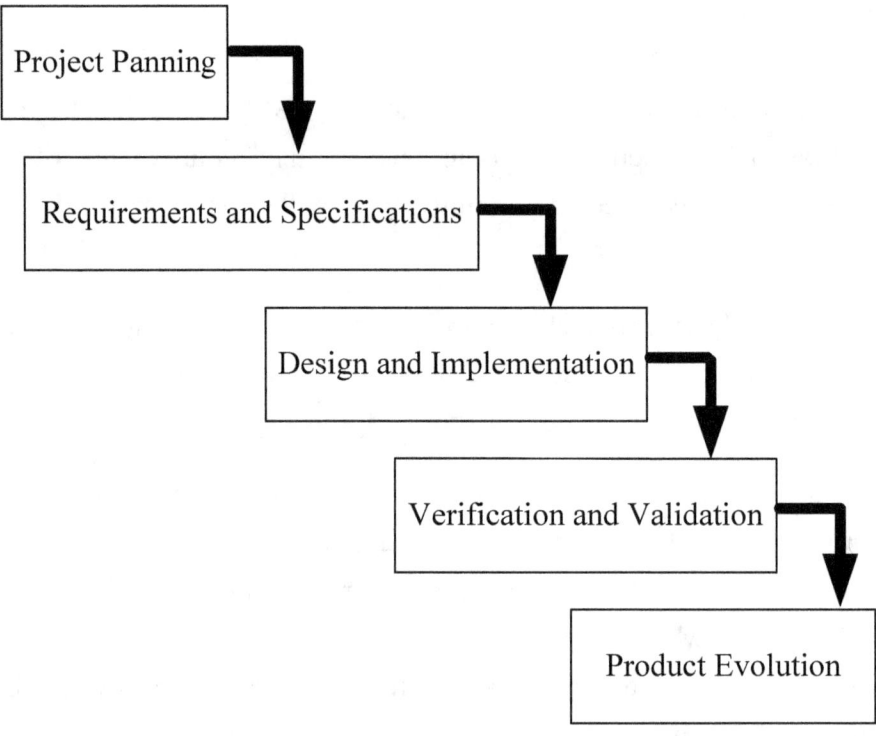

Figure 1-1   Five Phases of the Systems Development Life Cycle

1-1-1 Project Planning

Project planning determines the general goals of the systems development project. These general goals include: project scope determination; selection of the system process model; selection of the system engineering development technology; estimating applicable resources; determining system metric methodology; cost estimation; risk management; project scheduling and tracking; determining the configuration management approach; understanding the level of quality management; choosing system engineering tools; drawing up contracts; and determining post-project follow up.

1-1-2 Requirements and Specifications

The requirements and specifications phase consists of determining what the customer really requires. Requirements and specifications appertain to the problem space. When working on requirements and specifications, we usually only specify what the system is, but never think about how the system shall be manufactured.

1-1-3 Design and Implementation

The design and implementation phase belongs to the solution space. In other words, design and implementation try to secure a solution to meet or exceed customer requirements. It is opposite to requirements and specifications, design and implementation mainly consider how to manufacture the system, but not to specify what the system is.

1-1-4 Verification and Validation

The fourth step is called the verification and validation, abbreviated as V&V, phase. Verification uses proving technology. Validation uses testing technology. After the system product has been manufactured, we use either verification or validation to determine if or not the system product meets the requirements and specifications initially settled.

1-1-5 Product Evolution

Product evolution is the fifth, also the last, phase of system development life cycle. After verification and validation, we hand over the system product for the customer to use. Uses for several years, several month or even several days later, if has the necessity to carry on the next edition, either perceive that some part of wrong, some parts need the reinforcement, either the customer thinks that some places must change the requirements and specifications, even overhauls greatly, then must carry on the product evolution in accordance.

1-2 System Requirements Specification

System requirements specification (SyRS) is, in the systems development life cycle, a result of the requirements and specifications phase [Grad06, Lapl13]. That is, a system requirements specification is for the analysts to find out what the customers indeed expect the system to do for them. When working on the system requirements specification, we only specify what this system is, but never ask how this system shall be manufactured.

During the requirements and specifications phase, both customers and analysts need to coordinate closely, exchange the opinion fully, finally achieve the specifications document output, as shown in Figure 1-2.

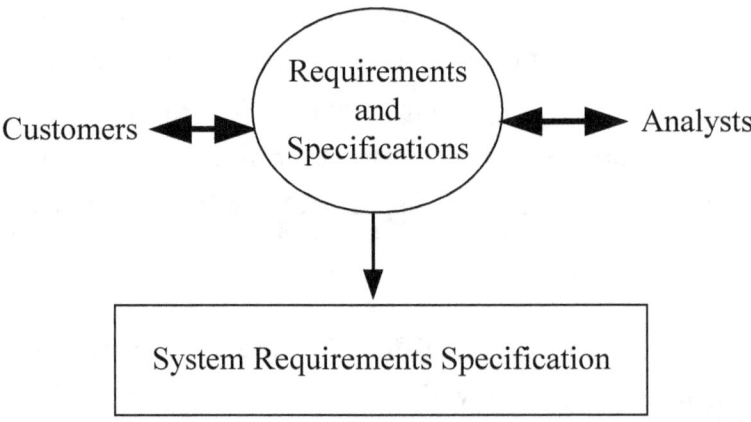

Figure 1-2   Work of Requirements and Specifications

Customers maintain great responsibility in the requirements and specifications work. The opinion and request on the system provided by the customer can be the very precious primitive information in the system requirements specification. Therefore, customers need to coordinate closely with analysts during the work of requirements and specifications.

Analysts also uphold magnificent responsibility in the requirements and specifications work. The analyst must be able to grasp specialized knowledge on computer hardware, firmware and software, is good at carrying on the abstract logical thinking and the creative thinking, can listen attentively to others' opinion. In addition, the analyst also is liable to assemble the system specifications document.

1-3 Multiple Views of a System

In general, a system is extremely complex that it consists of multiple views such as structure view, behavior view, function view, data view as shown in Figure 1-3 [Denn08, Kend10, Pres09, Somm06].

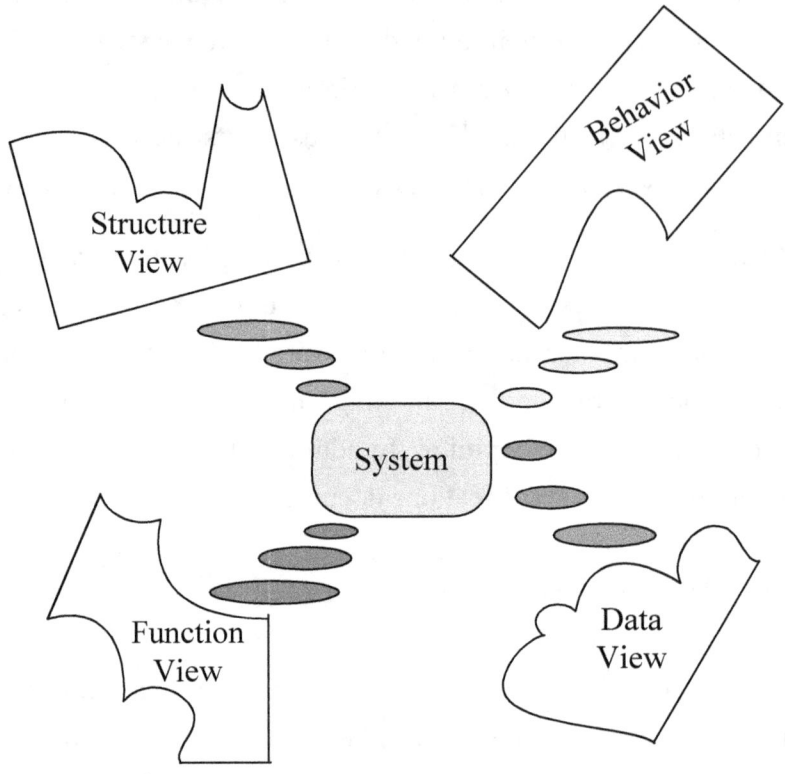

Figure 1-3   Multiple Views of a System

Among the above multiple views, the structure and behavior views are perceived as the two prominent ones. The structure view focuses on the systems structure which is described by components and their composition while the behavior view concentrates on the systems behavior which involves interactions [Chao15a, Chao15b, Chao15c, Chao15d, Chao15e, Hoar85, Miln89, Miln99] among the external environment's actors and components. Function and data views are considered to be other views as shown in Figure 1-4.

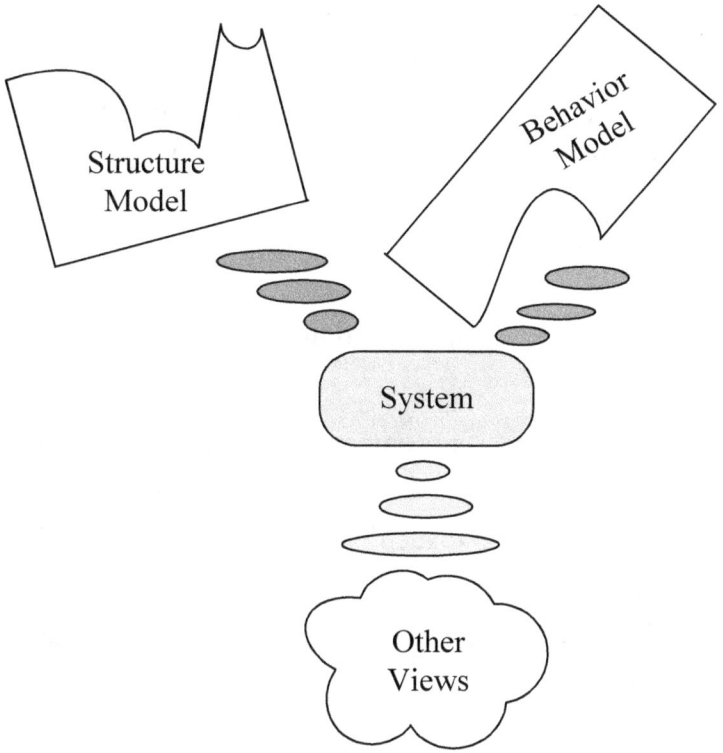

Figure 1-4   Structure, Behavior and Other Views

Either Figure 1-3 or Figure 1-4 represents the multiple views of a system. In some situations Figure 1-3 is used and in other situations Figure 1-4 is used.

Accordingly, a system is specified in Figure 1-5 to be an integrated whole of that system's multiple views, i.e., structure, behavior and other views, embodied in its assembled components, their interactions [Chao15a, Chao15b, Chao15c, Chao15d, Chao15e, Hoar85, Miln89, Miln99] with each other and the environment. Components are sometimes labeled as non-aggregated systems, parts, entities, objects and building blocks [Chao14a, Chao14b, Chao14c].

A system is an integrated whole of that system's multiple views, i.e., structure, behavior, and other views, embodied in its assembled components, their interactions with each other and the environment.

Figure 1-5   Specification of a System

Since multiple views are embodied in a system's assembled components which belong to the systems structure, they shall not exist alone. Multiple views must be loaded on the systems structure just like a cargo is loaded on a ship as shown in Figure 1-6. There will be no multiple views if there is no systems structure. Stand-alone multiple views are not meaningful.

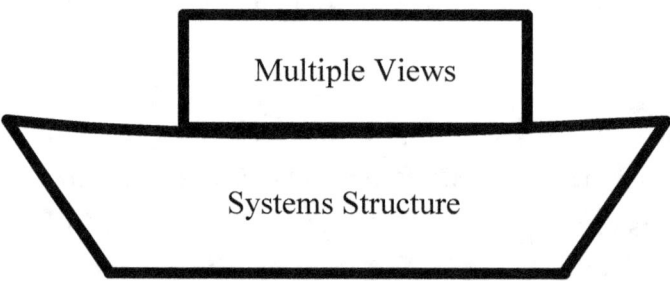

Figure 1-6   Multiple Views Loaded on the Systems Structure

1-4 Multiple Views Non-Integrated Approaches for Systems Requirements Specification 1.0

When specifying a system, the multiple views non-integrated approach, also known as the model multiplicity approach [Dori95, Dori02, Dori16], respectively picks a model for each view as shown in Figure 1-7, the structure view has the structure model; the behavior view has the behavior model; the function view has the function model; the data view has the data model. These multiple models, are heterogeneous and not related to each other, and thus become the primary cause of model multiplicity problems [Dori95, Dori02, Dori16, Pele02, Sode03].

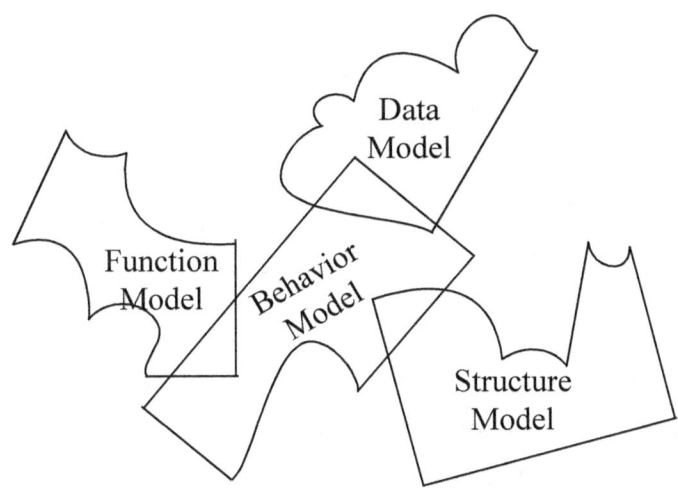

Figure 1-7   Multiple Views Non-Integrated Approach

Multiple views non-integrated approaches for system requirements specification (SyRS) 1.0 fall into four general categories: data-oriented, function-oriented, control-oriented and object-oriented [Grad13, Hatl00], as shown in Figure 1-8. Each of these approaches, more or less, fails to describe a system as an integrated whole of that system's multiple views.

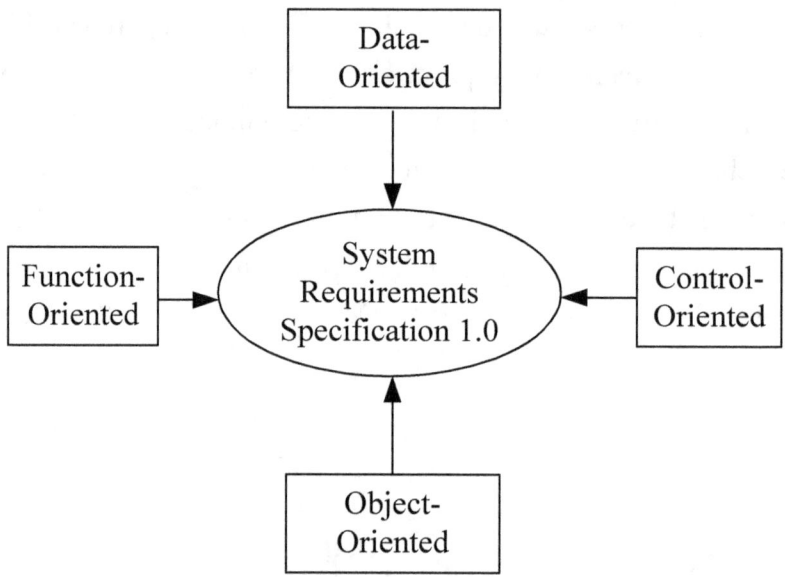

Figure 1-8   Multiple Views Non-Integrated Approaches
for System Requirements Specification (SyRS) 1.0

Data-oriented approaches for system requirements specification (SyRS) 1.0 stress the system state as a data structure. Jackson System Development (JSD) [Came89] and Entity Relationship Modeling (ERM) [Chen76] are primarily data-oriented. Data-oriented approaches concentrate only on data and completely neglect to integrate the systems structure and systems behavior. Therefore, data-oriented approaches are multiple views non-integrated and will never become an ideal SyRS approach.

Function-oriented   approaches   for   system   requirements specification (SyRS) 1.0 take the primary view of the way a system transforms input data into output data. Each transformation from input data into output data demonstrates a function of the system. A system may contain many such kinds of functions which represent the function view of the system. Classical Structured Analysis (SA) [DeMa79] fits into the category of process-based approaches, as do Structured Analysis and Design Technique (SADT) [Marc88] and Structured Systems Analysis and Design Method (SSADM) [Ashw90]. Process-based approaches concentrate only on the function view and completely neglect to integrate the systems structure and systems behavior. Just like data-oriented approaches, function-oriented approaches are multiple views non-integrated and will never become an ideal SyRS approach.

Control-oriented approaches for system requirements specification (SyRS) 1.0 emphasize synchronization, deadlock, exclusion, concurrency and process activation of a system. Petri Net [Reis92] and Flowcharting [Bash86] are primarily control-oriented. Control-oriented approaches concentrate only on the control view and completely neglect an integrated structure and behavior views which grasps the essential properties of a system. Just like data-oriented and function-oriented approaches, control-oriented approaches are multiple views non-integrated and will never become an ideal SyRS approach.

Object-oriented approaches for system requirements specification (SyRS) 1.0 specify the system as classes of objects and their behaviors. Object-oriented Analysis (OOA) [Booc07], fitting into the category of object-oriented approaches, looks at the problem domain, with the aim of producing a conceptual model of the information that exists in the area being analyzed. The result of object-oriented analysis is a description of what the system is behaviorally required to do, in the form of a conceptual model. That will typically be presented as a set of use cases and a number of activity diagrams. Object-oriented approaches stress both the structure view and the behavior view, but not an integrated structure and behavior views. Object-oriented approaches do not emphasize to integrate the systems structure and systems behavior. Like data-oriented, function-oriented and control-oriented approaches, object-oriented approaches are multiple views non-integrated and will never become an ideal SyRS approach.

1-5 Multiple Views Integrated Approaches for Systems Requirements Specification 2.0

When specifying a system, the multiple views integrated approach, also known as the model singularity approach [Dori95, Dori02, Dori16, Pele02, Sode03], instead of picking many heterogeneous and unrelated models, will use only one single model as shown in Figure 1-9. The structure, behavior, function and data views are all integrated in this one single model which represents an integrated whole of that system's multiple views [Chao14a, Chao14b, Chao14c].

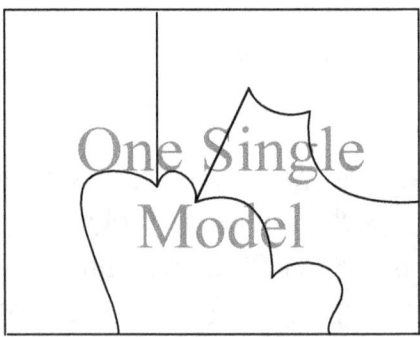

Figure 1-9   Multiple Views Integrated Approach

Multiple views integrated approaches for system requirements specification (SyRS) 2.0 specify a system as an integrated whole of that system's multiple views.

Chapter 2: Systems Structure and Systems Behavior

Systems structure and systems behavior are the two most significant views of a system. Systems structure, specified by components, their operations and their composition, refers to the type of connection between the components of a system. Systems behavior, specified by the interactions between and among the components and environment, refers to the interconnectivities a system in conjunction with its environment.

2-1 Structure of Systems

Every system forms a whole. In general, structure of systems is the type of connection between the components of a system. More specifically, we specify the structure of a system by 1) components, 2) their operations and 3) their composition.

Components are something relatively indivisible in one system [Hoff10, Shel11]. For example, *Parking_Garages_CityMap_UI*, *Inquire_Parking_Fees_UI*, *Pay_Parking_Fees_UI*, *Parking_Starting_Time_Daemon*, *Parking_End_Time_Daemon*, *SPCASIS_Database*, *Driver_GPS_P (P = AAA0000 to ZZZ9999)*, *Parking_Starting_Time_Sensor_Q (Q = 000 to 999)* and *Parking_End_Time_Sensor_R (R = 000 to 999)* are components of the *Smart Parking Cloud Applications and Services IoT System* (SPCASIS) as shown in Figure 2-1.

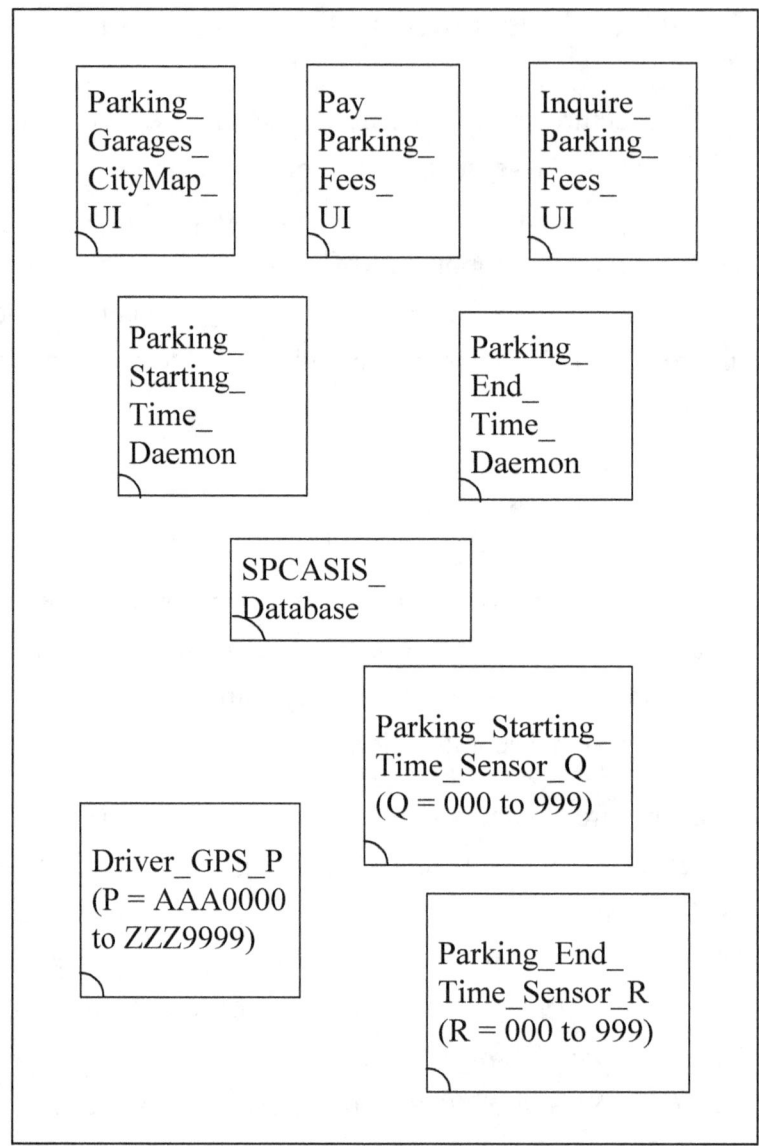

Figure 2-1   Components of the
Smart Parking Cloud Applications and Services IoT System

An operation provided by each component represents a procedure
or method or function of the component [Chao14a, Chao14b, Chao14c].

Each component in a system must possess at least one operation. Figure 2-2 shows the operations of all components of the *Smart Parking Cloud Applications and Services IoT System* (SPCASIS). In the figure, component *Parking_Garages_CityMap_UI* has two operations: *Show_Parking_Garages_CityMap* and *Reserve_One_Parking_Slot*; component *Inquire_Parking_Fees_UI* has one operation: *Show_Parking_Fees*; component *Pay_Parking_Fees_UI* has one operation: *Pay_Parking_Fees*; component *Parking_Starting_Time_Daemon* has one operation: *Fork_PSTD_Process*; component *Parking_End_Time_Daemon* has one operation: *Fork_PETD_Process*; component *SPCASIS_Database* has six operations: *SQL_Select_Parking_Garages*, *SQL_Insert_One_Parking_Slot*, *SQL_Insert_Parking_Starting_Time*, *SQL_Select_Parking_Duration*, *SQL_Insert_Parking_Fees_Payment* and *SQL_Insert_Parking_End_Time*; component *Driver_GPS_P* *(P = AAA0000 to ZZZ9999)* has one operation: *Driver_GPS_Positioning*; component *Parking_Starting_Time_Sensor_Q (Q = 000 to 999)* has two operations: *Sense_Parking_Starting_Time* and *Return_Parking_Starting_Time*; component *Parking_End_Time_Sensor_R (R = 000 to 999)* has two operations: *Sense_Parking_End_Time* and *Return_Parking_End_Time*.

32

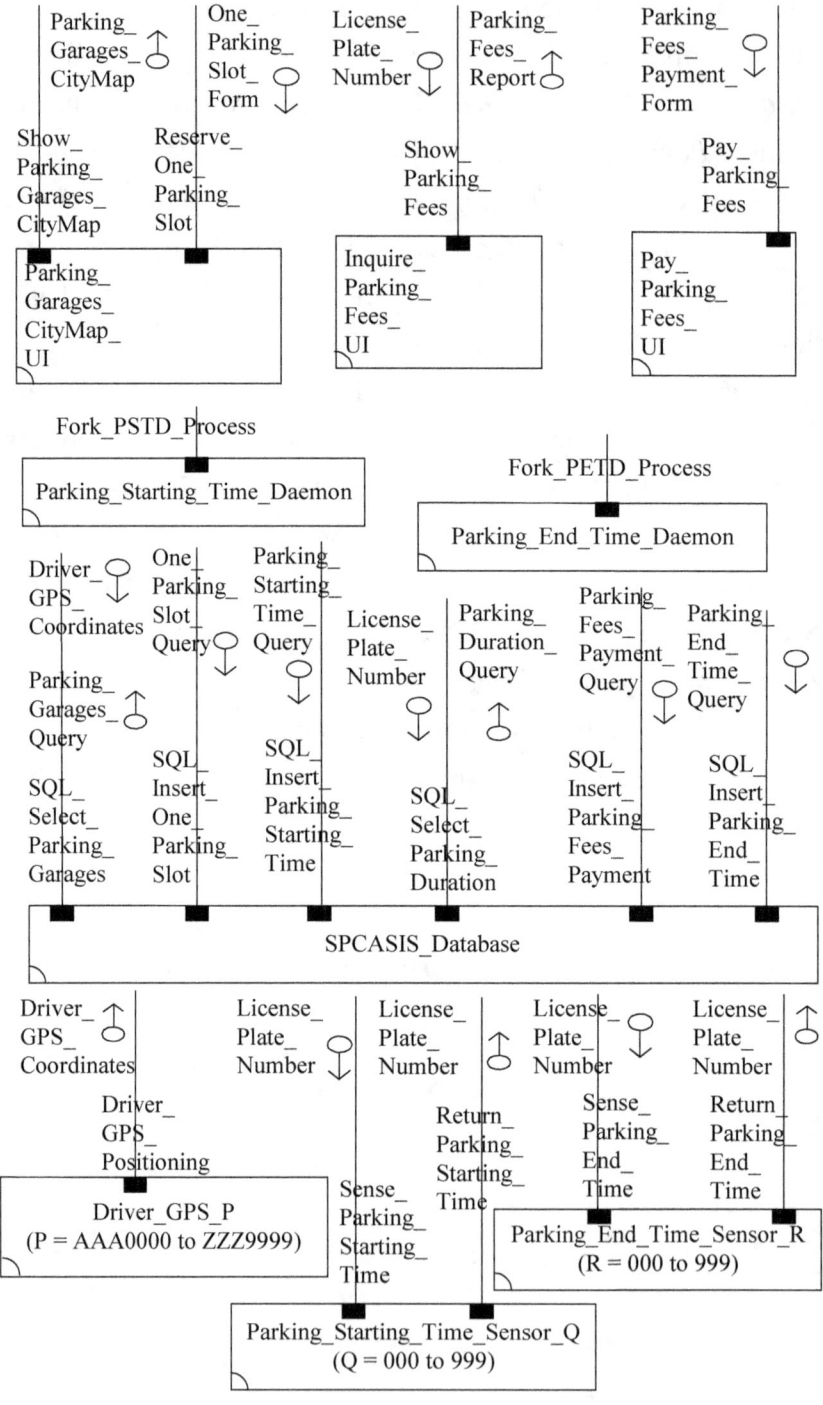

Figure 2-2   Operations of all Components of
the *Smart Parking Cloud Applications and Services IoT System*

Composition of components designs the structural composition and decomposition of a system. For example, Figure 2-3 shows that, in the *Smart Tourism City Cloud Applications and Services IoT System* (STCCASIS), *SPCASIS* is composed of *Application_Layer*, *Data_Layer* and *Technology_Layer*; *Application_Layer* is composed of *Presentation_Layer* and *Logic_Layer*; *Presentation_Layer* is composed of *Parking_Garages_CityMap_UI*, *Inquire_Parking_Fees_UI* and *Pay_Parking_Fees_UI*; *Logic_Layer* is composed of *Parking_Starting_Time_Daemon* and *Parking_End_Time_Daemon*; *Data_Layer* is composed of *SPCASIS_Database*; *Technology_Layer* is composed of *Driver_GPS_P (P = AAA0000 to ZZZ9999)*, *Parking_Starting_Time_Sensor_Q (Q = 000 to 999)* and *Parking_End_Time_Sensor_R (R = 000 to 999)*.

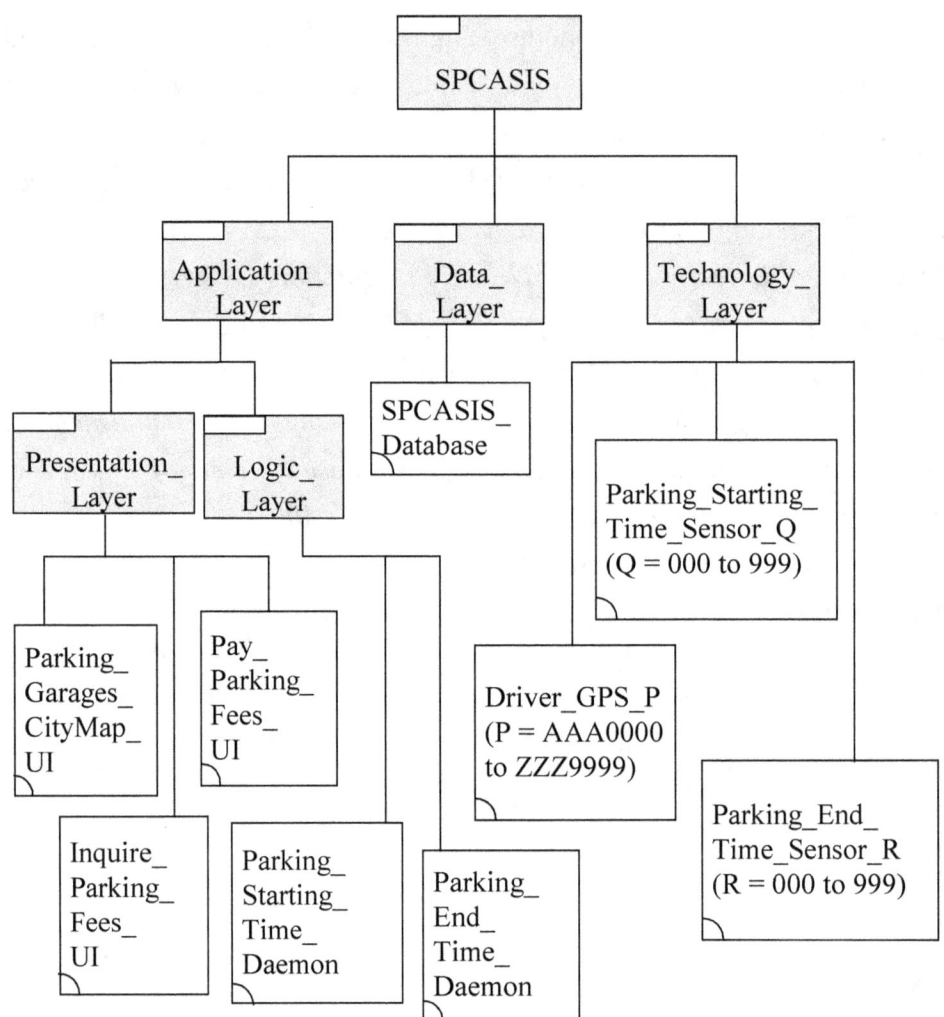

Figure 2-3   Structural Composition of
the *Smart Parking Cloud Applications and Services IoT System*

2-2 Behavior of Systems

Systems behavior refers to the interactions a system in conjunction with its environment. It is the response of a system to various stimuli, whether internal or external, conscious or subconscious, overt or covert, and voluntary or involuntary.

For example, Figure 2-4 demonstrates five individual behaviors: *Finding_and_Reserving_a_Vacant_Parking_Slot*, *Sensing_Parking_Starting_Time*, *Inquiring_Parking_Fees*, *Paying_Parking_Fees*, *Sensing_Parking_End_Time* that refer to the interactions the the *Smart Parking Cloud Applications and Services IoT System* in conjunction with its environment.

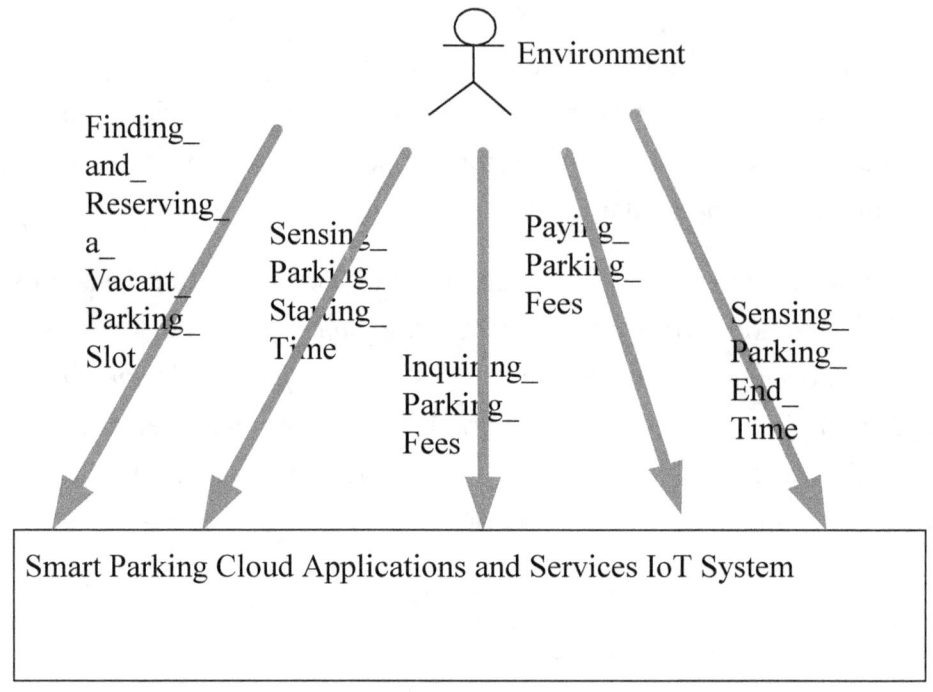

Figure 2-4   Behaviors of
the *Smart Parking Cloud Applications and Services IoT System*

For each behavior, the environment always initiates the interaction and will lead more follow-up interactions to be realized among components. For example, Figure 2-5 demonstrates that interactions between and among the environment and the *Parking_Garages_CityMap_UI*, *SPCASIS_Database* and *Driver_GPS_P(P = AAA0000 to ZZZ9999)* components shall draw forth the *Finding_and_Reserving_a_Vacant_Parking_Slot* behavior.

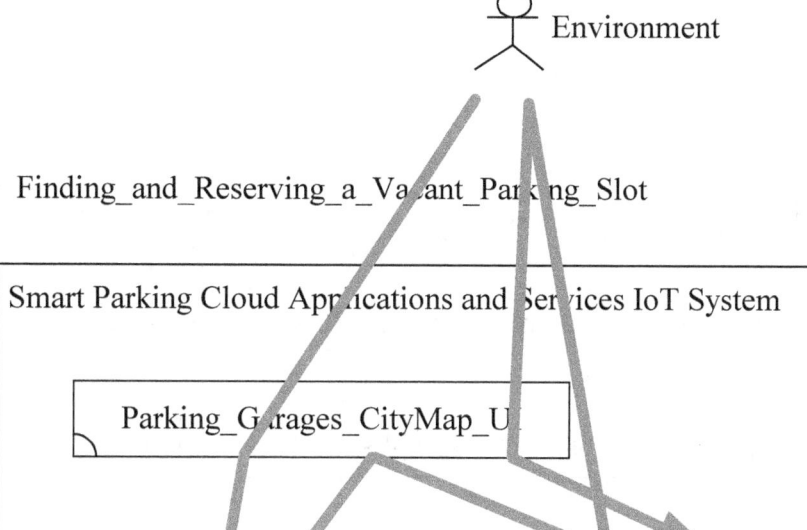

Figure 2-5   Interactions that Draw forth
the *Finding_and_Reserving_a_Vacant_Parking_Slot* Behavior

As a second example, Figure 2-6 demonstrates that interactions between and among the environment and the *Parking_Starting_Time_Daemon*, *SPCASIS_Database* and *Parking_Starting_Time_Sensor_Q (Q = 000 to 999)* components shall draw forth the *Sensing_Parking_Starting_Time* behavior.

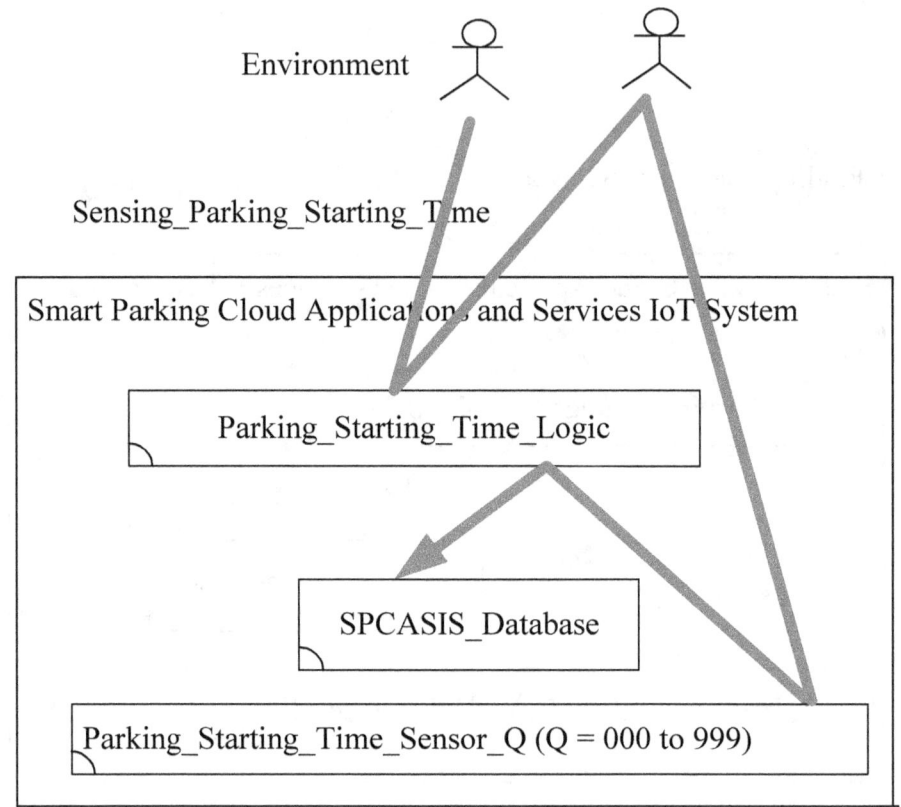

Figure 2-6   Interactions that Draw forth
the *Sensing_Parking_Starting_Time* Behavior

As a third example, Figure 2-7 demonstrates that interactions between and among the environment and the *Inquire_Parking_Fees_UI* and *SPCASIS_Database* components shall draw forth the *Inquiring_Parking_Fees* behavior.

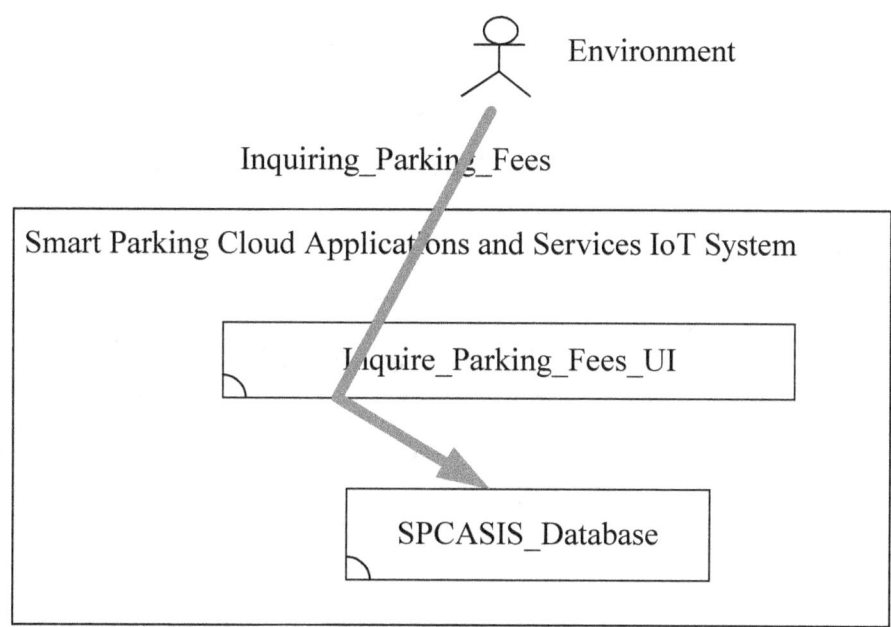

Figure 2-7   Interactions that Draw forth
the *SInquiring_Parking_Fees* Behavior

As a fourth example, Figure 2-8 demonstrates that interactions between and among the environment and the *Pay_Parking_Fees_UI* and *SPCASIS_Database* components shall draw forth the *Inquiring_Parking_Fees* behavior.

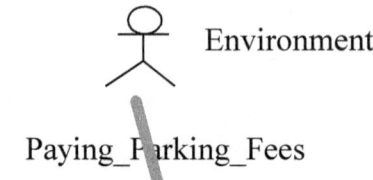

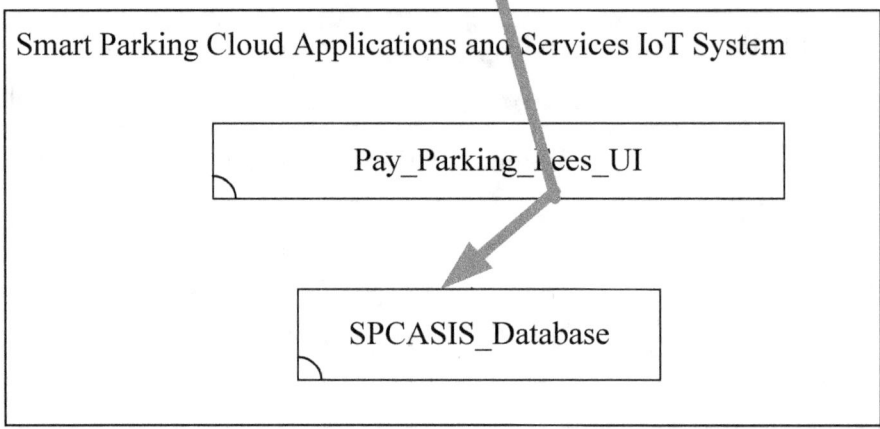

Figure 2-8   Interactions that Draw forth
the *Paying_Parking_Fees* Behavior

As a fifth example, Figure 2-9 demonstrates that interactions between and among the environment and the *Parking_Starting_Time_Daemon*, *SPCASIS_Database* and *Parking_End_Time_Sensor_R (R = 000 to 999)* components shall draw forth the *Sensing_Parking_End_Time* behavior.

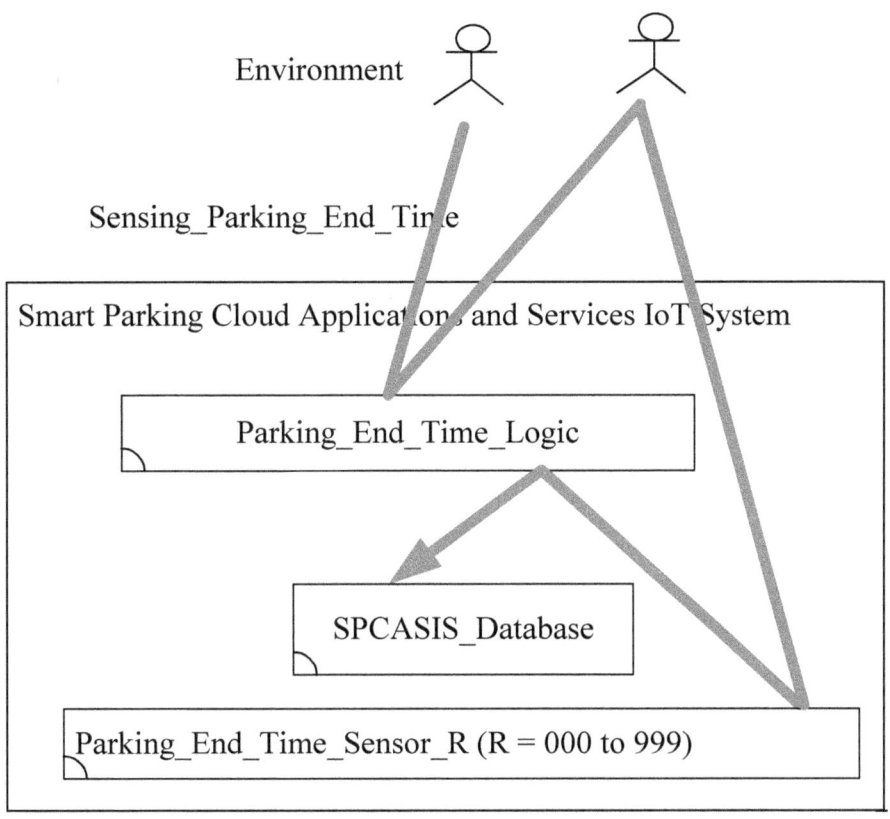

Figure 2-9   Interactions that Draw forth
the *Sensing_Parking_End_Time* Behavior

Chapter 3: Structure-Behavior Coalescence

A system has been specified hopefully to be an integrated whole, embodied in its assembled components, their interactions with each other and the environment. Since systems structure and systems behavior are the two most prominent views of a system, integrating the systems structure and systems behavior apparently is the best way to achieve a truly integrated whole of a system. Because system requirements specification 1.0 does not specify the integration of systems structure and systems behavior, very likely it will never be able to actually form an integrated whole of a system.

Structure-behavior coalescence (SBC) provides an elegant way to integrate the systems structure and systems behavior, and hence achieves a truly integrated whole, of a system. A truly integrated whole sets a path to achieve the desired system requirements specification (SyRS). SBC facilitates an integrated whole. Therefore, we conclude that SBC sets a path to achieve the system requirements specification. System requirements specification 2.0 uses the SBC approach and is highly adequate in specifying a system.

3-1 Integrated Whole to Achieve the System Requirements Specification

A system has been specified hopefully to be an integrated whole, embodied in its assembled components, their interactions with each other and the environment. In other words, an integrated whole sets a path to achieve the system requirements specification (SyRS) as shown in Figure 3-1.

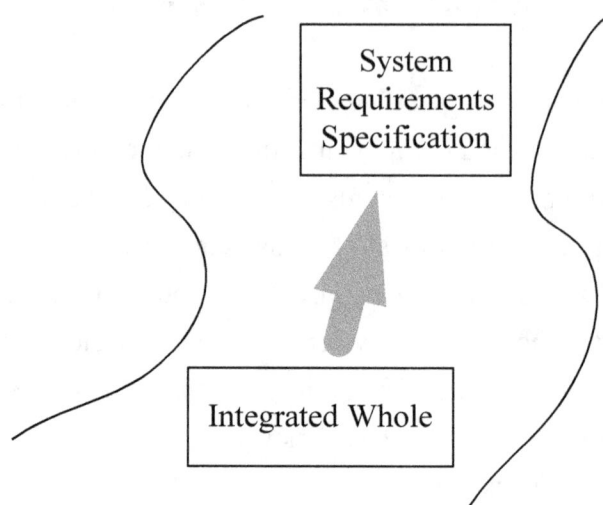

Figure 3-1   Integrated Whole to Achieve
the System Requirements Specification

In one system requirements specification, different systems structures may draw forth the same integrated whole as shown in Figure 3-2.

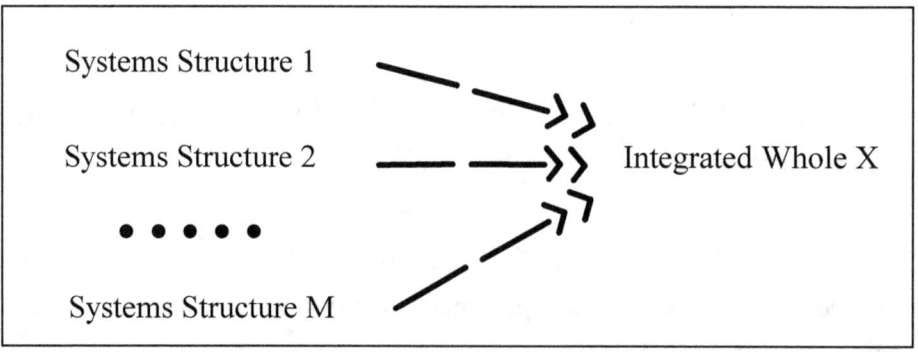

Figure 3-2   Different Systems Structures Draw Forth
the Same Integrated Whole

Since there is only one systems structure exists in one system requirements specification, one systems structure will draw forth one integrated whole as shown in Figure 3-3.

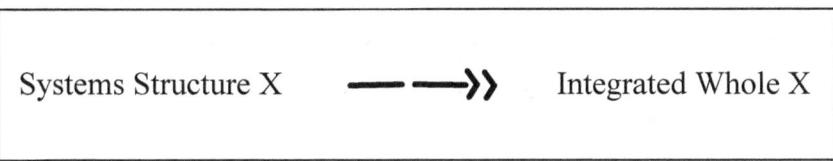

Systems Structure X   — — —⟩⟩   Integrated Whole X

Figure 3-3   One Systems Structure Draws Forth
One Integrated Whole

We conclude that in one system requirements specification, an integrated whole must be attached to or built on a systems structure. In other words, an integrated whole shall not exist alone; it must be loaded on a systems structure just like a cargo is loaded on a ship as shown in Figure 3-4. There will be no integrated whole if there is no systems structure. A stand-alone integrated whole with no systems structure is not meaningful.

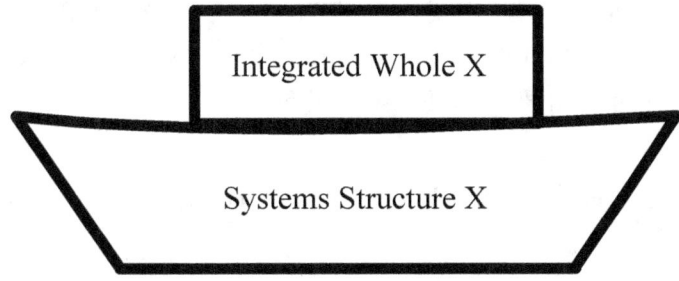

Figure 3-4   An Integrated Whole Must be Loaded on
a Systems Structure

3-2 Integrating the Systems Structure and Systems Behavior

By integrating the systems structures and systems behaviors, we obtain structure-behavior coalescence (SBC) within a system. Since systems structures and systems behaviors are so tightly integrated, we sometimes claim that the core theme of structure-behavior coalescence is: "Systems Architecture = Systems Structure + Systems Behavior," as shown in Figure 3-5.

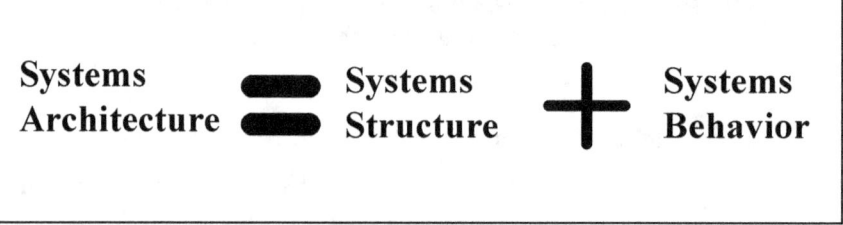

Figure 3-5   Core Theme of Structure-Behavior Coalescence

So far, integrating the systems structure and systems behavior has never been proposed or suggested besides the SBC approach. In most cases, systems behaviors are separated from systems structures when specifying a system [Hoff10, Pres09, Shel11, Somm06].

3-3 Structure-Behavior Coalescence to Facilitate an Integrated Whole

Since systems structure and systems behavior are the two most prominent views of a system, integrating the systems structure and systems behavior apparently is the best way to achieve a truly integrated whole of a system. If we are not able to integrate the systems structure and systems behavior, then there is no way that we are able to integrate the whole system. In other words, structure-behavior coalescence (SBC) facilitates a truly integrated whole as shown in Figure 3-6.

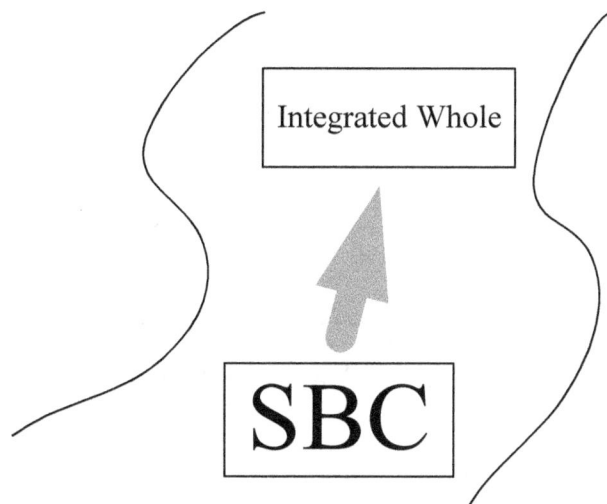

Figure 3-6   SBC Facilitates an Integrated Whole

Since system requirements specification 1.0 does not specify the integration of systems structure and systems behavior, very likely it will never be able to actually form an integrated whole of a system. In this situation, system requirements specification 1.0 is powerless in specifying a system adequately.

3-4 Structure-Behavior Coalescence to Achieve the System Requirements Specification

Figure 3-1 declares that an integrated whole sets a path to achieve the desired system requirements specification. Figure 3-6 declares that structure-behavior coalescence facilitates a truly integrated whole.

Combining the above two declarations, we conclude that the structure-behavior coalescence (SBC) approach sets a path to achieve the system requirements specification as shown in Figure 3-7.

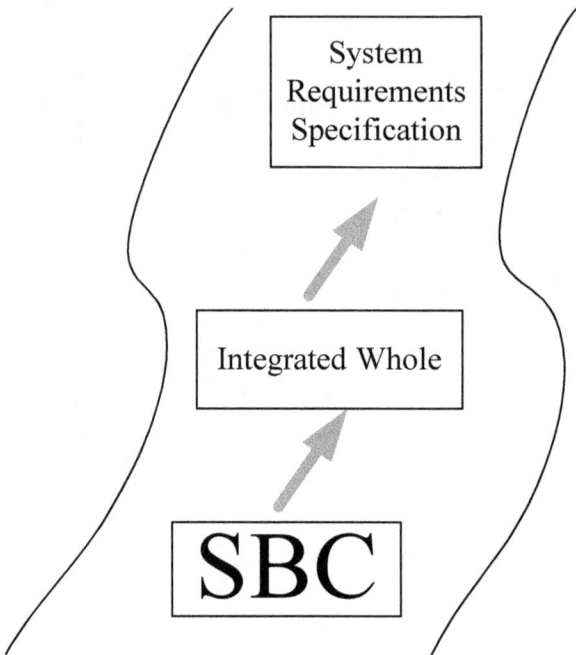

Figure 3-7   SBC to Achieve
the System Requirements Specification

In the SBC approach, different systems structures may draw forth the same systems behavior as shown in Figure 3-8.

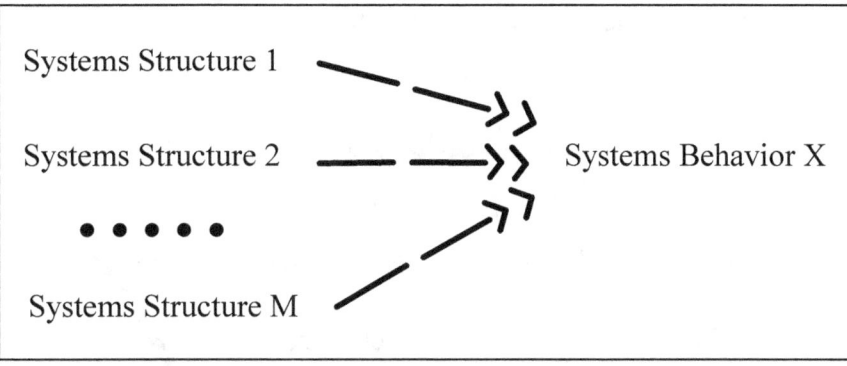

Figure 3-8 Different Systems Structures Draw Forth
the Same Systems Behavior

Since there is only one systems structure exists in one system requirements specification, one systems behavior will always be attached to or built on one systems structure as shown in Figure 3-9.

Systems Structure X ——⟶⟩⟩ Systems Behavior X

Figure 3-9 One Systems Behavior is Attached to
One Systems Structure

We conclude that in the SBC approach, a systems behavior must be attached to or built on a systems structure. In other words, a systems behavior can not exist alone; it must be loaded on a systems structure just like a cargo is loaded on a ship as shown in Figure 3-10. There will be no systems behavior if there is no systems structure. A stand-alone systems behavior with no systems structure is not meaningful.

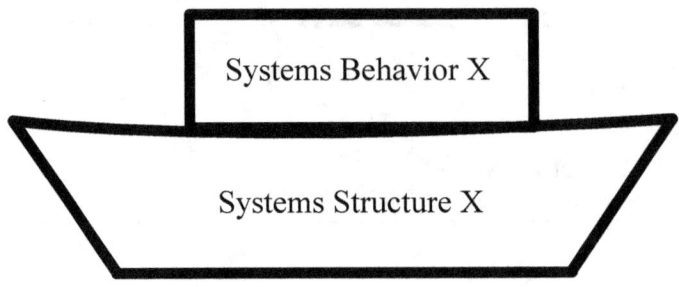

Figure 3-10   A Systems Behavior Must be Loaded on
a Systems Structure

3-5 SBC Approach for System Requirements Specification 2.0

Since structure-behavior coalescence (SBC) provides an elegant way to integrate the systems structure and systems behavior, we shall include it in the specification of a system. Figure 3-11 shows how the system requirements specification 2.0 specifies a system.

A system,
through the SBC approach,
truly is an integrated whole,
embodied in its assembled components,
their interactions with each other and the environment.

Figure 3-11   System Requirements Specification 2.0
Specifying a System

A system specified by the system requirements specification 2.0 has the following characteristics: 1) it emphasizes the system's structure-behavior coalescence; 2) it is a truly integrated whole; 3) it is embodied in its assembled components; 4) components are interacting (or handshaking) [Chao15a, Chao15b, Chao15c, Chao15d, Chao15e, Hoar85, Miln89, Miln99] with each other and the environment; and 5) it uses structural decomposition [Chao14a, Chao14b, Chao14c, Ghar11] rather than functional decomposition [Scho10].

Structure-behavior coalescence (SBC) provides an elegant way to integrate the systems structure and systems behavior of a system. System requirements specification 2.0 uses the SBC approach to formally specify the integration of systems structure and systems behavior of a system. System requirements specification 2.0 contains three fundamental diagrams: a) architecture hierarchy diagram, b) component operation diagram and c) interaction flow diagram.

So far, we have introduced the system requirements specification 2.0 which should be able to appropriately specify a system. In the following chapters, we shall elaborate the details of the system requirements specification 2.0.

3-6 SBC Model Singularity

Channel-Based Single-Queue SBC Process Algebra (C-S-SBC-PA) [Chao17a], Channel-Based Multi-Queue SBC Process Algebra (C-M-SBC-PA) [Chao17b], Channel-Based Infinite-Queue SBC Process Algebra (C-I-SBC-PA) [Chao17c], Operation-Based Single-Queue SBC Process Algebra (O-S-SBC-PA) [Chao17d], Operation-Based Multi-Queue SBC Process Algebra (O-M-SBC-PA) [Chao17e] and Operation-Based Infinite-Queue SBC Process Algebra (O-I-SBC-PA) [Chao17f] are the six specialized SBC process algebras. The SBC process algebra

(SBC-PA) shown in Figure 3-12 is a model singularity approach.

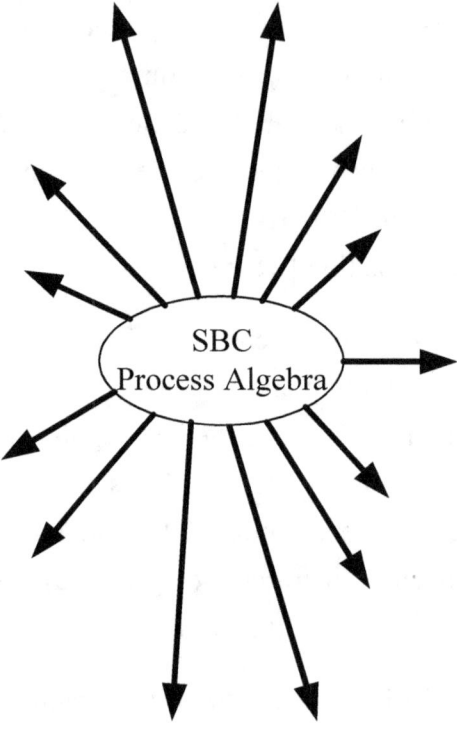

Figure 3-12   SBC-PA is a Model Singularity Approach.

The system requirements specification (SyRS) 2.0 is also a model singularity approach. With SBC mind set sitting in the kernel, the SyRS 2.0 single model shown in Figure 3-13 is therefore able to represent all structural views such as architecture hierarchy diagram (AHD), component operation diagram (COD), and behavioral views such as interaction flow diagram (IFD).

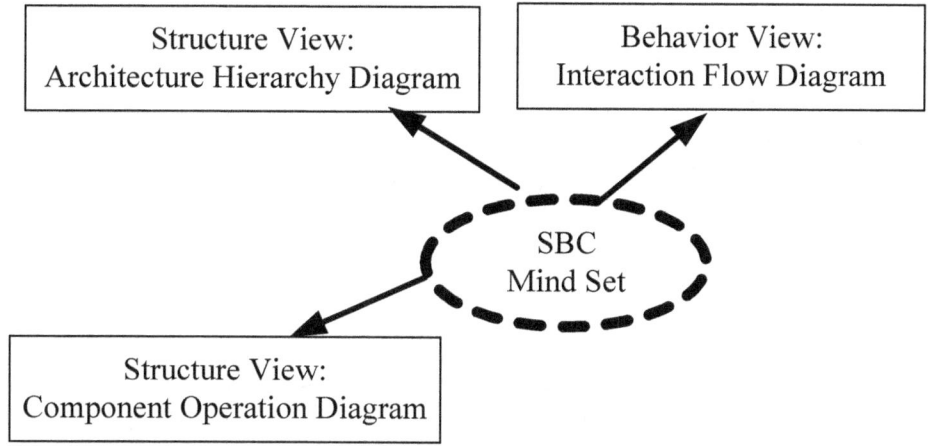

Figure 3-13   SyRS 2.0 is a Model Singularity Approach.

The combination of SBC process algebra (SBC-PA) and system requirements specification (SyRS) 2.0 is shown in Figure 3-14, again as a model singularity approach.

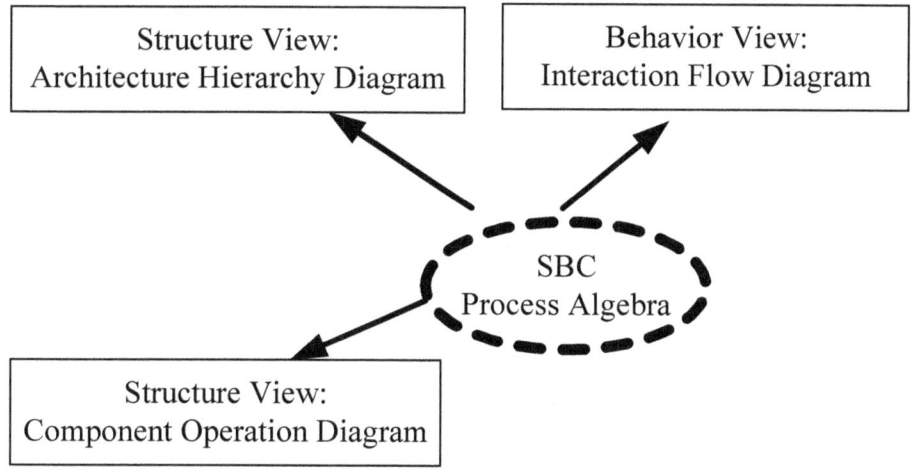

Figure 3-14   SBC Model is a Model Singularity Approach.

PART II: SBC APPROACH FOR SYSTEM REQUIREMENTS SPECIFICATIONS 2.0

Chapter 4: Architecture Hierarchy Diagram

SBC approach for system requirements specification (SyRS) 2.0 uses an architecture hierarchy diagram (AHD) to specify the multi-level decomposition and composition of a system.

4-1 Decomposition and Composition

The following is an example of systems decomposition and composition. The *SPCASIS* is composed of *Application_Layer*, *Data_Layer* and *Technology_Layer* as shown in Figure 4-1. *Application_Layer*, *Data_Layer* and *Technology_Layer* are subsystems comprising the *SPCASIS*.

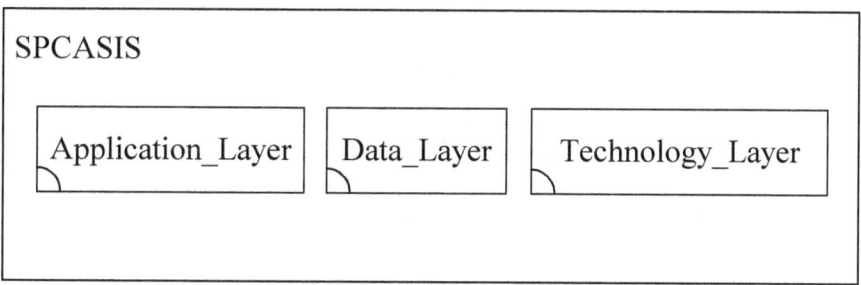

Figure 4-1   Decomposition and Composition of
the *SPCASIS*

Another example indicates that the *LPRCASIS* is composed of *Application_Layer*, *Data_Layer* and *Technology_Layer* as shown in Figure 4-2. *Application_Layer*, *Data_Layer* and *Technology_Layer* are subsystems comprising the *LPRCASIS*.

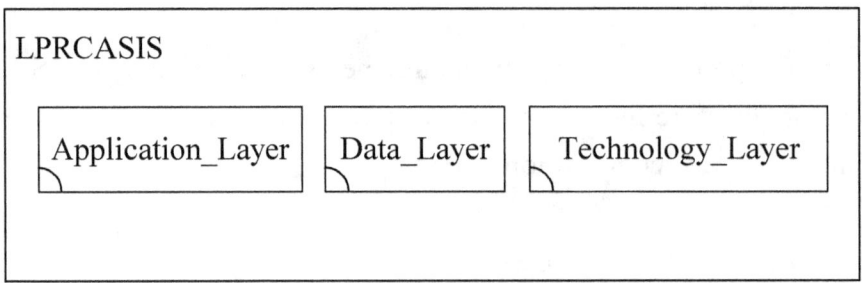

Figure 4-2   Decomposition and Composition of
the *LPRCASIS*

The architecture hierarchy diagram (AHD) is used to represent the decomposition and composition of a software system. As an example, an AHD of the *SPCASIS* is shown in Figure 4-3.

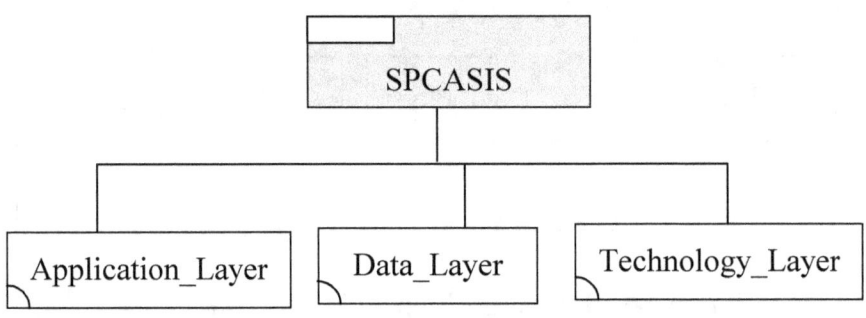

Figure 4-3   AHD of the *SPCASIS*

As a second example, Figure 4-4 shows an AHD of the *LPRCASIS*.

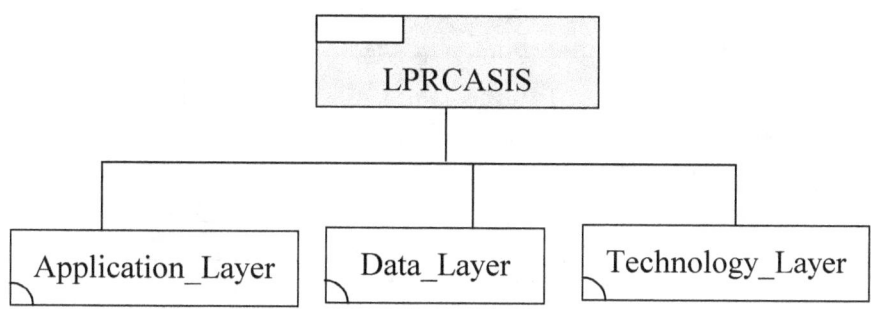

Figure 4-4   AHD of the *LPRCASIS*

4-2 Multi-Level Decomposition and Composition

The subsystem may also contain subsystems as we further decompose it. For example, *Application_Layer* is a subsystem of the *SPCASIS*, and we can further decompose it into *Parking_Garages_CityMap_UI*, *Inquire_Parking_Fees_UI*, *Pay_Parking_Fees_UI*, *Parking_Starting_Time_Daemon*, *Parking_End_Time_Daemon*; *Data_Layer* is also a subsystem of the *SPCASIS*, and we can further decompose it into *SPCASIS_Database*; *Technology_Layer* is also a subsystem of the *SPCASIS*, and we can further decompose it into *Parking_Starting_Time_Sensor_Q (Q = 000 to 999), Parking_End_Time_Sensor_R (R = 000 to 999)* as shown in Figure 4-5.

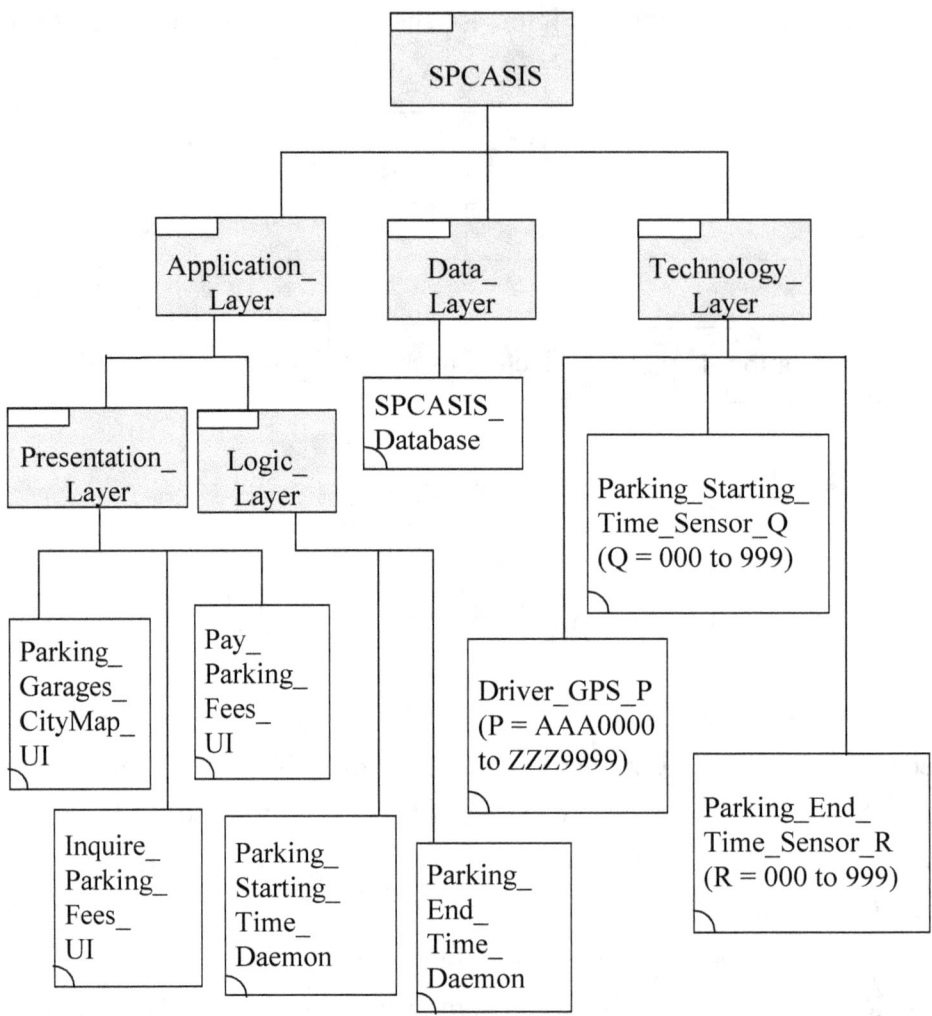

Figure 4-5   Multi-Level De/Composition of
the *SPCASIS*

As a second example, *Application_Layer* is a subsystem of the *LPRCASIS*, and we can further decompose it into *Occurring_Landslide_UI*, *Alerts_Notifying_UI*, *Emergency_Response_Starting_Time_UI*, *Emergency_Response_End_Time_UI*, *Landslide_Signs_Daemon*; *Data_Layer* is also a subsystem of the *LPRCASIS*, and we can further decompose it into *LPRCASIS_Database*; *Technology_Layer* is also a subsystem of the *SPCASIS*, and we can further decompose it into *Landslide_Signs_Sensor_N (N = A0000 to Z9999)* as shown in Figure 4-6.

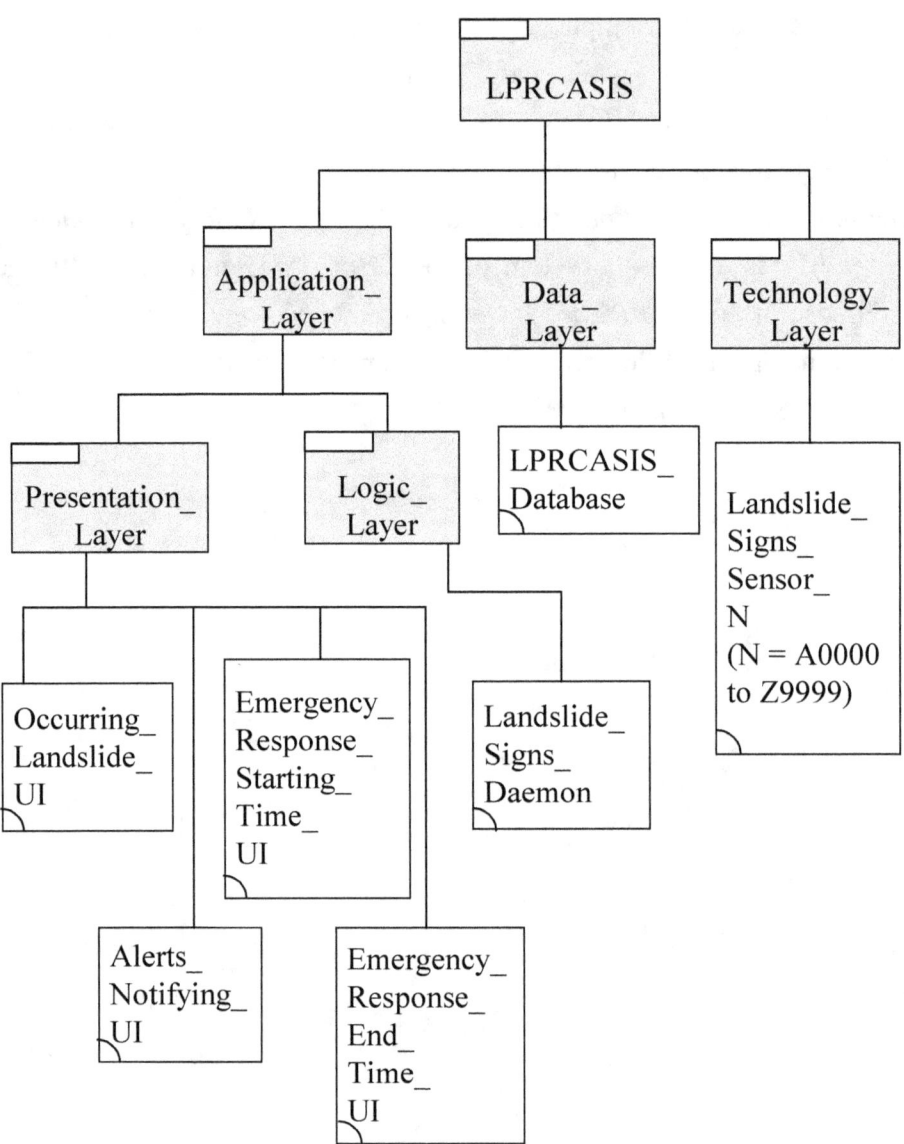

Figure 4-6   Multi-Level De/Composition of
the *LPRCASIS*

4-3 Aggregated and Non-Aggregated Systems

Any system (at any level) involved with multi-level decomposition and composition of a system is either aggregated or non-aggregated. The definition of aggregated and non-aggregated systems is shown in Figure 4-7.

Definition of Aggregated Systems:

A system (within an AHD) is aggregated if it is composed of any sub-system.

Definition of Non-aggregated Systems

A system (within an AHD) is non-aggregated if it is NOT composed of any sub-system.

Figure 4-7   Definition of Aggregated and
Non-aggregated Systems

Non-aggregated systems are sometimes referred to as components, parts, entities, objects and building blocks [Chao14a, Chao14b, Chao14c].

In the multi-level (hierarchical) decomposition and composition, any system is either aggregated or non-aggregated, but not both. For example, in Figure 4-3, *Application_Layer* is a non-aggregated system, not an aggregated system. As an interesting contrast, in Figure 4-5, *Application_Layer* is an aggregated system, not a non-aggregated system.

As a second example, in Figure 4-4, *Technology_Layer* is a non-aggregated system, not an aggregated system. As an interesting contrast, in Figure 4-6, *Technology_Layer* is an aggregated system, not a non-aggregated system.

Chapter 5: Component Operation Diagram

SBC approach for system requirements specification (SyRS) 2.0 uses a component operation diagram (COD) to specify all components' operations of a system.

5-1 Operations of Each Component

An operation provided by each component represents a procedure or method or function of the component. If other components request this component to perform an operation, then shall use it to accomplish the operation request.

Each component in a system must possess at least one operation. A component should not exist in a system if it does not possess any operation. Figure 5-1 shows that the *Parking_Starting_Time_Sensor_Q (Q = 000 to 999)* component has two operations: *Sense_Parking_Starting_Time* and *Return_Parking_Starting_Time*.

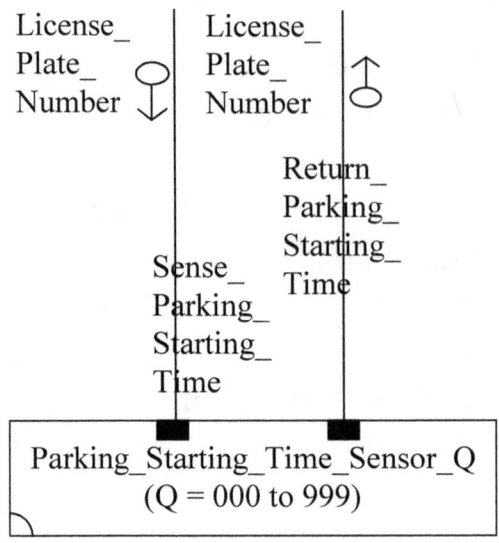

Figure 5-1   Two Operations of the
Parking_Starting_Time_Sensor_Q (Q = 000 to 999) Component

An operation formula is utilized to fully represent an operation. An operation formula includes a) operation name, b) input parameters and c) output parameters as shown in Figure 5-2.

$$Operation\_Name \ (In \ i_1, i_2, ..., i_m \ ; Out \ o_1, o_2, ..., o_n \ )$$

Figure 5-2   Operation Formula

Operation name is the name of this operation. In a system, every operation name should be unique. Duplicate operation names shall not be allowed in any system.

An operation may have several input and output parameters. The input and output parameters, gathered from all operations, represent the input data and output data views of a system [Date03, Elma10]. As shown in Figure 5-3, component *Parking_Starting_Time_Sensor_Q (Q = 000 to 999)* possesses the *Sense_Parking_Starting_Time* operation which has the *License_Plate_Number* input parameter (with the arrow direction pointing to the component); component *Parking_Starting_Time_Sensor_Q (Q = 000 to 999)* also possesses the *Return_Parking_Starting_Time* operation which has the *License_Plate_Number* output parameter (with the arrow direction opposite to the component).

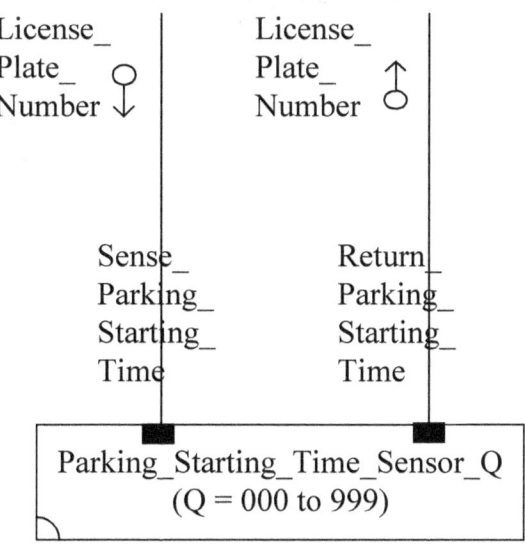

Figure 5-3   Input/Output Parameters

Data formats of input and output parameters can be described by data type specifications. There are two sets of data types: primitive and composite [Date03, Elma10]. Figure 5-4 shows the primitive data type specification of the *License_Plate_Number* parameter occurring in the *Show_Parking_Fees(In License_Plate_Number; Out Parking_Fees_Report)*, *SQL_Select_Parking_Duration(In License_Plate_Number; Out Parking_Duration_Query)*, *Sense_Parking_Starting_Time(In License_Plate_Number)*, *Return_Parking_Starting_Time(Out License_Plate_Number)*, *Sense_Parking_End_Time(In License_Plate_Number)* and *Return_Parking_End_Time(Out License_Plate_Number)* operation formulas.

Parameter	Data Type	Instances
License_ Plate_ Number	Text	ADA3456

Figure 5-4   Primitive Data Type Specification

Figure 5-5 shows the composite data type specification of the *Parking_Garages_Query* output parameter occurring in the *SQL_Select_Parking_Garages(In Driver_GPS_Coordinates; Out Parking_Garages_Query)* operation formula.

Parameter	*Parking_Garages_Query*
Data Type	TABLE of Parking_Garage: Text Parking_Garage_GPS_Coordinates: Text Available_Slots: Integer End TABLE ;
Instances	

Parking_ Garage	Parking_ Garage_ GPS_ Coordinates	Available_ Slots
Ampco System Parking	34.039311, -118.438579	400
O K Parking	34.03919, -118.441495	298
Ace Parking	34.035844 , -118.444571	366

Figure 5-5   Composite Data Type Specification of *Parking_Garages_Query*

Figure 5-6 shows the composite data type specification of the *One_Parking_Slot_Query* input parameter occurring in the *SQL_Insert_One_Parking_Slot(In One_Parking_Slot_Query)* operation formula.

Parameter	*One_Parking_Slot_Query*

Data Type	TABLE of
	  Parking_Garage: Text
	  Parking_Slot: Text
	  License_Plate_Number: Text
	  Reservation_Date_Time: Text
	End TABLE ;

Instances	

Parking_ Garage	Parking_ Slot	License_ Plate_ Number	Reservation_ Date_ Time
Ampco System Parking	7899	ADA3456	20150828, 08:50:30

Figure 5-6   Composite Data Type Specification of *One_Parking_Slot_Query*

5-2 Drawing the Component Operation Diagram

For a system, COD is used to specify all components' operations. Figure 5-7 shows the *Smart Parking Cloud Applications and Services IoT System's COD*. In the figure, component *Parking_Garages_CityMap_UI* has two operations: *Show_Parking_Garages_CityMap* and *Reserve_One_Parking_Slot*; component *Inquire_Parking_Fees_UI* has one operation: *Show_Parking_Fees*; component *Pay_Parking_Fees_UI* has one operation: *Pay_Parking_Fees*; component *Parking_Starting_Time_Daemon* has one operation:

Fork_PSTD_Process; component *Parking_End_Time_Daemon* has one operation: *Fork_PETD_Process*; component *SPCASIS_Database* has six operations: *SQL_Select_Parking_Garages*, *SQL_Insert_One_Parking_Slot*, *SQL_Insert_Parking_Starting_Time*, *SQL_Select_Parking_Duration*, *SQL_Insert_Parking_Fees_Payment* and *SQL_Insert_Parking_End_Time*; component *Driver_GPS_P (P = AAA0000 to ZZZ9999)* has one operation: *Driver_GPS_Positioning*; component *Parking_Starting_Time_Sensor_Q (Q = 000 to 999)* has two operations: *Sense_Parking_Starting_Time* and *Return_Parking_Starting_Time*; component *Parking_End_Time_Sensor_R (R = 000 to 999)* has two operations: *Sense_Parking_End_Time* and *Return_Parking_End_Time*.

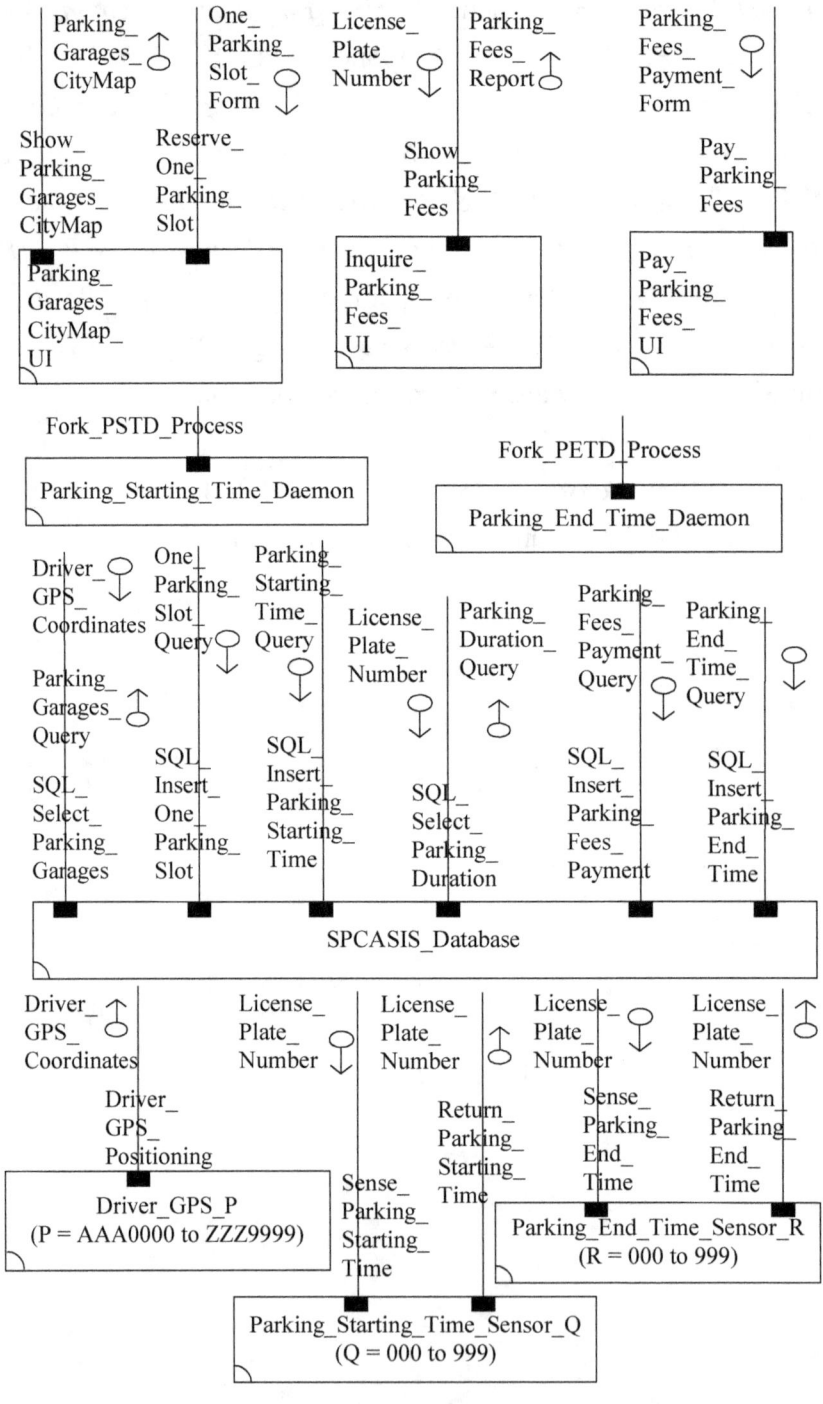

Figure 5-7   COD of the
Smart Parking Cloud Applications and Services IoT System

Chapter 6: Interaction Flow Diagram

SBC approach for systems requirements specification (SyRS) 2.0 uses an interaction flow diagram (IFD) to specify each individual behavior of the overall behavior of a system.

6-1 Individual Behavior Represented by Interaction Flow Diagram

The overall behavior of a system consists of many individual behaviors. Each individual behavior represents an execution path. An IFD is utilized to specify such an individual behavior.

Figure 6-1 demonstrates that the *Smart Parking Cloud Applications and Services IoT System* (SPCASIS) has five behaviors; thus, it has five IFDs.

System	IFD
SPCASIS	Finding_and_Reserving_a_Vacant_Parking_Slot
	Sensing_Parking_Starting_Time
	Inquiring_Parking_Fees
	Paying_Parking_Fees
	Sensing_Parking_End_Time

Figure 6-1 *Smart Parking Cloud Applications and Services IoT System* has Five IFDs

Figure 6-2 demonstrates that the *Landslide Prevention and Relief Cloud Applications and Services IoT System* (LPRCASIS) has five behaviors; thus, it has five IFDs.

System	IFD
LPRCASIS	Sensing_Landslide_Signs
	Recording_Occurring_Landslide
	Alerts_Notifying
	Recording_Emergency_Response_Starting_Time
	Recording_Emergency_Response_End_Time

Figure 6-2 *Landslide Prevention and Relief Cloud Applications and Services IoT System* has Five IFDs

6-2 Drawing the Interaction Flow Diagram

Let us now explain the usage of interaction flow diagram (IFD) by drawing an IFD step by step. Figure 6-3 demonstrates an IFD of the *Paying_Parking_Fees* behavior. The X-axis direction is from the left side to right side and the Y-axis direction is from the above to the below. Inside an IFD, there are four elements: a) external environment's actor, b) components, c) interactions and d) input/output parameters. Participants of the interaction, such as the external environment's actor and each component, are laid aside along the X-axis direction on the top of the diagram. The external environment's actor which initiates the sequential interactions is always placed on the most left side of the X-axis. Then, interactions among the external environment's actor and components successively in turn decorate along the Y-axis direction. The first interaction is placed on the top of the Y-axis position. The last interaction is placed on the bottom of the Y-axis position. Each interaction may carry several input and/or output parameters.

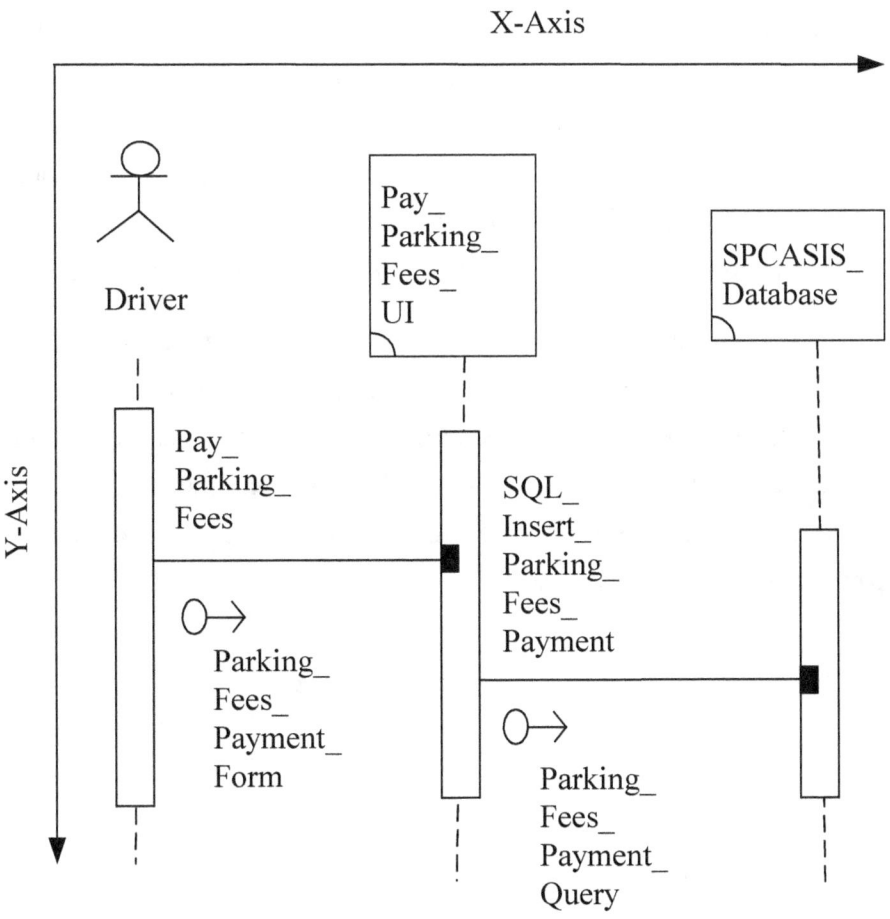

Figure 6-3   IFD of the *Paying_Parking_Fees* Behavior

In Figure 6-3, *Driver* is an external environment's actor. *Pay_Parking_Fees_UI* and *SPCASIS_Database* are components. *Pay_Parking_Fees* is an operation, carrying the *Parking_Fees_Payment_Form* input parameter, which is provided by the *Pay_Parking_Fees_UI* component. *SQL_Insert_Parking_Fees_Payment* is an operation, carrying the *Parking_Fees_Payment_Query* input parameter, which is provided by the *SPCASIS_Database* component.

The execution path of Figure 6-3 is as follows. First, actor *Driver* interacts with the *Pay_Parking_Fees_UI* component through the *Pay_Parking_Fees* operation call interaction, carrying the *Parking_Fees_Payment_Form* input parameters. Finally, component *Pay_Parking_Fees_UI* interacts with the *SPCASIS_Database* component through the *SQL_Insert_Parking_Fees_Payment* operation call interaction, carrying the *Parking_Fees_Payment_Query* input parameter.

For each interaction, the solid line stands for operation call while the dashed line stands for operation return. The operation call and operation return interactions, if using the same operation name, belong to the identical operation. Figure 6-4 exhibits two interactions (operation call interaction and operation return interaction) having the identical "*Show_Parking_Fees*" operation.

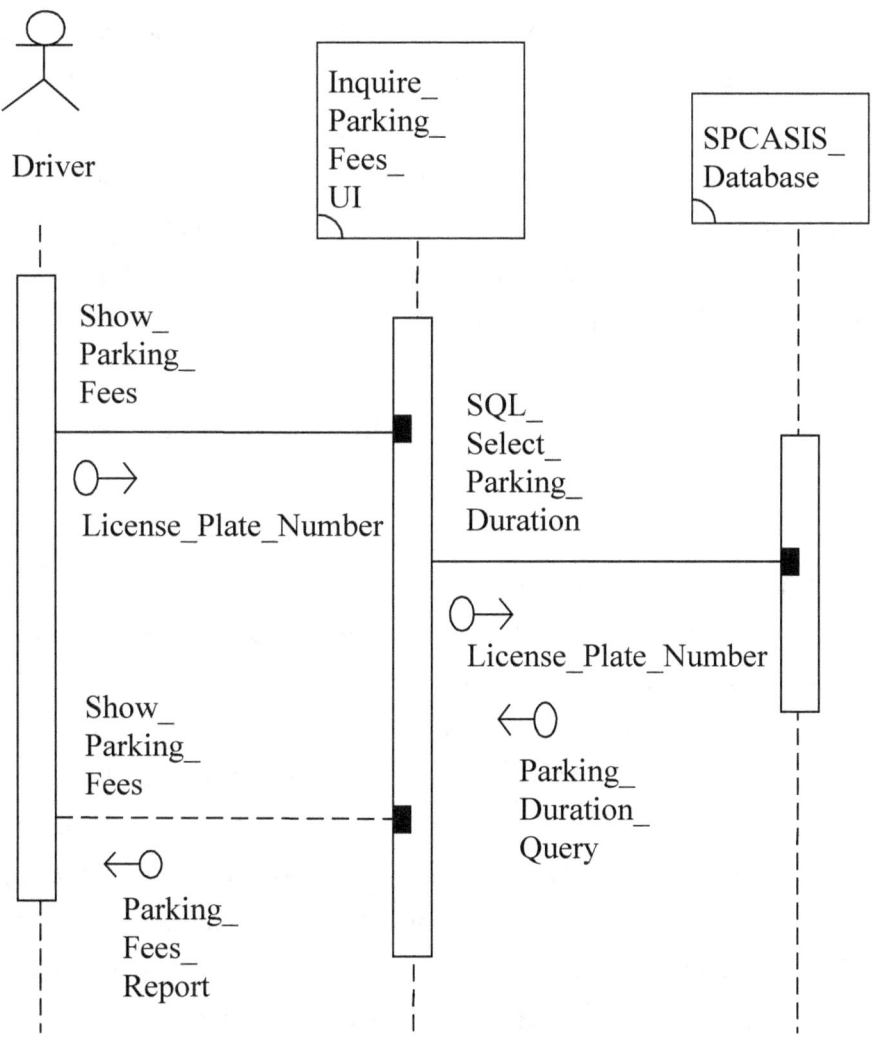

Figure 6-4   Two Interactions Have the Identical Operation

The execution path of Figure 6-4 is as follows. First, actor *Driver* interacts with the *Inquire_Parking_Fees_UI* component through the *Show_Parking_Fees* operation call interaction, carrying the

License_Plate_Number input parameter. Next, component *Inquire_Parking_Fees_UI* interacts with the *SPCASIS_Database* component through the *SQL_Select_Parking_Duration* operation call interaction, carrying the *License_Plate_Number* input parameter and the *Parking_Duration_Query* output parameter. Finally, actor *Driver* interacts with the *Inquire_Parking_Fees_UI* component through the *Show_Parking_Fees* operation return interaction, carrying the *Parking_Fees_Report out*put parameter.

An interaction flow diagram may contain a conditional expression. Figure 6-5 shows such an example which has the following execution path. First, external environment's actor *Employee* interacts with the *Computer* component through the *Open* operation call interaction, carrying the *Task_No* input parameter. Next, if the *var_1 < 4 & var_2 > 7* condition is true then component *Computer* shall interact with the *Skype* component through the *Op_1* operation call interaction and component *Skype* shall interact with the *Earphone* component through the *Op_4* operation call interaction, carrying the *Skype_Earphone* output parameter; else if the *var_3 = 99* condition is true then component *Computer* shall interact with the *Skype* component through the *Op_2* operation call interaction and component *Skype* shall interact with the *Speaker* component through the *Op_5* operation call interaction, carrying the *Skype_Speaker* output parameter; else component *Computer* shall interact with the *Youtube* component through the *Op_3* operation call interaction and component *Youtube* shall interact with the *Speaker* component through the *Op_6* operation call interaction, carrying the *Youtube_Speaker* output parameter. Continuingly, if the *var_1 < 4 & var_2 > 7* condition is true then component *Computer* shall interact with the *Skype* component through the *Op_1* operation return interaction, carrying the *Status_1* output parameter; else if the *var_3 = 99* condition is true then component *Computer* shall interact with the *Skype* component through the *Op_2* operation return interaction, carrying the *Status_2* output parameter; else component *Computer* shall interact with the *Youtube* component through the *Op_3* operation return interaction, carrying the *Status_3* output parameter. Finally, external environment's actor *Employee* interacts with the *Computer* component through the *Open* operation return interaction, carrying the *Status* output parameter.

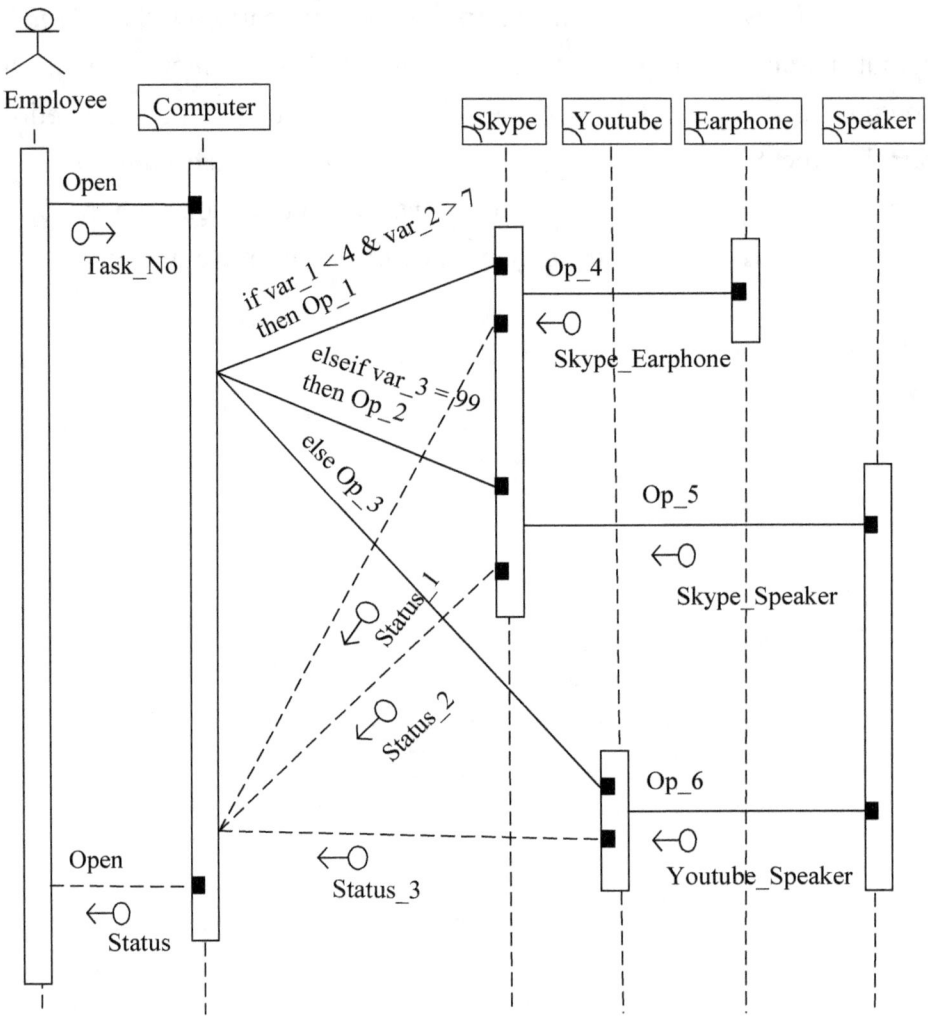

Figure 6-5   Conditional Interaction

Several Boolean conditions are shown in Figure 6-5. They are"*var_1 < 4 & var_2 > 7*" and "*var_3 = 99*". Variables, such as *var_1*, *var_2* and *var_3*, appearing in the Boolean condition can be local or global variables [Prat00, Seth96].

PART III: CASES STUDY

Chapter 7: System Requirement Specification 2.0 of the Smart Parking Cloud Applications and Services IoT System

As the numbers of vehicles on the road are increasing day by day parking problems are bound to exist. Parking is costly and limited in almost every major city in the world. A city consists of a group of parking garages. A parking garage consists of a group of parking slots. Searching for a vacant parking slot in a metropolitan area is difficult for most drivers. It commonly results more traffic congestion and air pollution by constantly cruising in certain area only for an available parking slot. For instance, a recent survey shows that during rush hours in most big cities, the traffic generated by cars searching for parking slots takes up to 40% of the total traffic. To alleviate such traffic congestion and improve the convenience for drivers, many smart parking systems aiming to satisfy the drivers as well as parking service providers have been deployed [Bi06].

A cloud applications and services IoT system is the network of physical things or objects embedded with electronics, software, sensors and connectivity to enable it to achieve greater value and service by exchanging data with the users, operators and other connected devices. Each thing or object is uniquely identifiable through its embedded cloud computing [Bern09] system but is able to interoperate within the existing internet infrastructure.

Advancements in cloud applications and services IoT systems present enormous potential for accurate monitoring and providing service to the customers and administrators. A cost effective solution to this service can be provided by wireless sensor networks which consists of larger number of sensors placed in existing parking garages without installing expensive cabling and are capable of adjusting with the cheap and easily available sensors. The information obtained from each sensor is processed collaboratively to evaluate meaning metrics such as parking-garage availability and checking-in and checking-out date and time of all vehicles. Smart parking cloud applications and services IoT system (SPCASIS) covers a great deal of parking garages. A parking garage consists of a group of parking slots.

Behaviors of SPCASIS consist of: a) behavior of *Finding_and_Reserving_a_Vacant_Parking_Slot*, b) behavior of *Sensing_Parking_Starting_Time*, c) behavior of *Inquiring_Parking_Fees*, d) behavior of *Paying_Parking_Fees* and e) behavior of *Sensing_Parking_End_Time*.

Using the structure-behavior coalescence (SBC) approach, we shall go through: a) architecture hierarchy diagram, b) component operation diagram and c) interaction flow diagram, to accomplish the system requirements specification (SyRS) 2.0 for SPCASIS.

7-1 Architecture Hierarchy Diagram

SyRS 2.0 uses an architecture hierarchy diagram (AHD) to specify the multi-level composition and decomposition of the *Smart Parking Cloud Applications and Services IoT System* (SPCASIS) as shown in Figure 7-1. In the figure, *SPCASIS* is composed of *Application_Layer*, *Data_Layer* and *Technology_Layer*; *Application_Layer* is composed of *Presentation_Layer* and *Logic_Layer*; *Presentation_Layer* is composed of *Parking_Garages_CityMap_UI*, *Inquire_Parking_Fees_UI* and *Pay_Parking_Fees_UI*; *Logic_Layer* is composed of *Parking_Starting_Time_Daemon* and *Parking_End_Time_Daemon*; *Data_Layer* is composed of *SPCASIS_Database*; *Technology_Layer* is composed of *Driver_GPS_P (P = AAA0000 to ZZZ9999)*,

Parking_Starting_Time_Sensor_Q (Q = 000 to 999) and
Parking_End_Time_Sensor_R (R = 000 to 999).

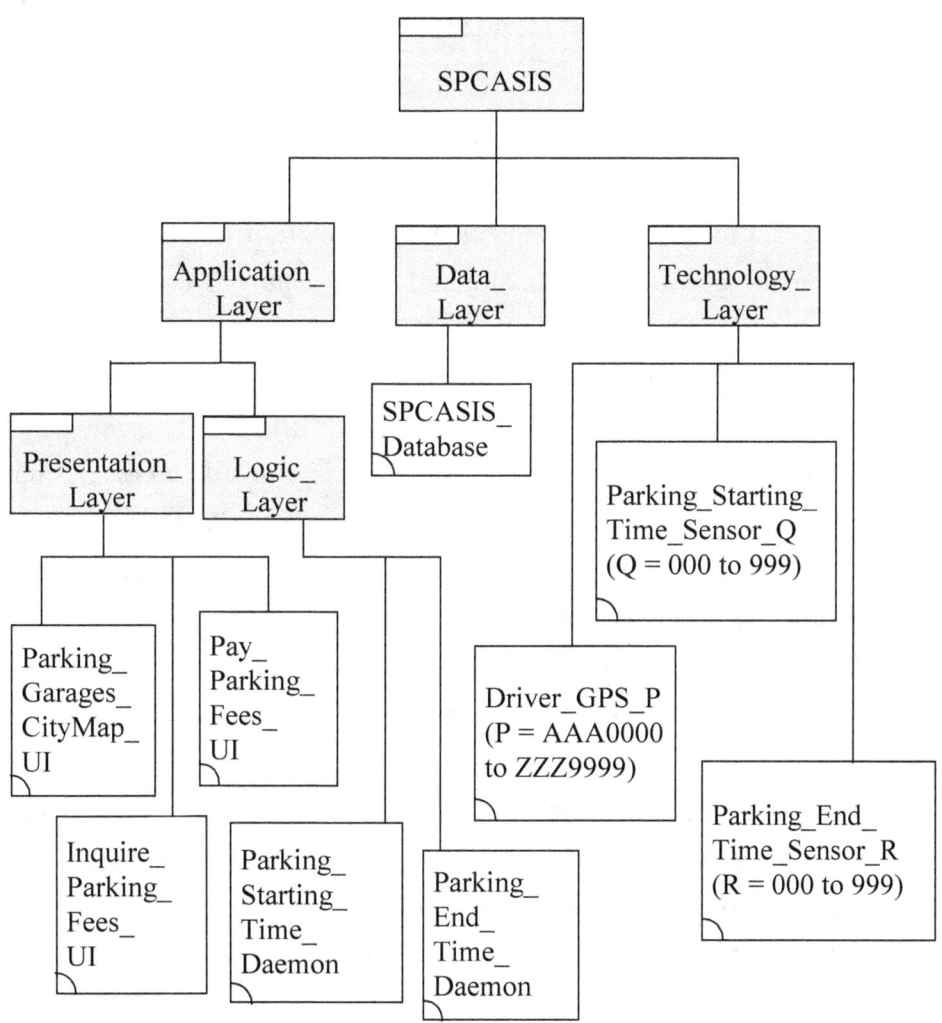

Figure 7-1   AHD of the *SPCASIS*

In Figure 7-1, *SPCASIS, Application_Layer, Presentation_Layer,
Logic_Layer, Data_Layer* and *Technology_Layer* are aggregated systems
while   *Parking_Garages_CityMap_UI,   Inquire_Parking_Fees_UI,
Pay_Parking_Fees_UI,   Parking_Starting_Time_Daemon,*

Parking_End_Time_Daemon, *SPCASIS_Database*, *Driver_GPS_P (P = AAA0000 to ZZZ9999)*, *Parking_Starting_Time_Sensor_Q (Q = 000 to 999)* and *Parking_End_Time_Sensor_R (R = 000 to 999)* are non-aggregated systems.

7-2 Component Operation Diagram

SyRS 2.0 uses a component operation diagram (COD) to specify the operations of all components of the *Smart Parking Cloud Applications and Services IoT System* (SPCASIS) as shown in Figure 7-2. In the figure, component *Parking_Garages_CityMap_UI* has two operations: *Show_Parking_Garages_CityMap* and *Reserve_One_Parking_Slot*; component *Inquire_Parking_Fees_UI* has one operation: *Show_Parking_Fees*; component *Pay_Parking_Fees_UI* has one operation: *Pay_Parking_Fees*; component *Parking_Starting_Time_Daemon* has one operation: *Fork_PSTD_Process*; component *Parking_End_Time_Daemon* has one operation: *Fork_PETD_Process*; component *SPCASIS_Database* has six operations: *SQL_Select_Parking_Garages*, *SQL_Insert_One_Parking_Slot*, *SQL_Insert_Parking_Starting_Time*, *SQL_Select_Parking_Duration*, *SQL_Insert_Parking_Fees_Payment* and *SQL_Insert_Parking_End_Time*; component *Driver_GPS_P (P = AAA0000 to ZZZ9999)* has one operation: *Driver_GPS_Positioning*; component *Parking_Starting_Time_Sensor_Q (Q = 000 to 999)* has two operations: *Sense_Parking_Starting_Time* and *Return_Parking_Starting_Time*; component *Parking_End_Time_Sensor_R (R = 000 to 999)* has two operations: *Sense_Parking_End_Time* and *Return_Parking_End_Time*.

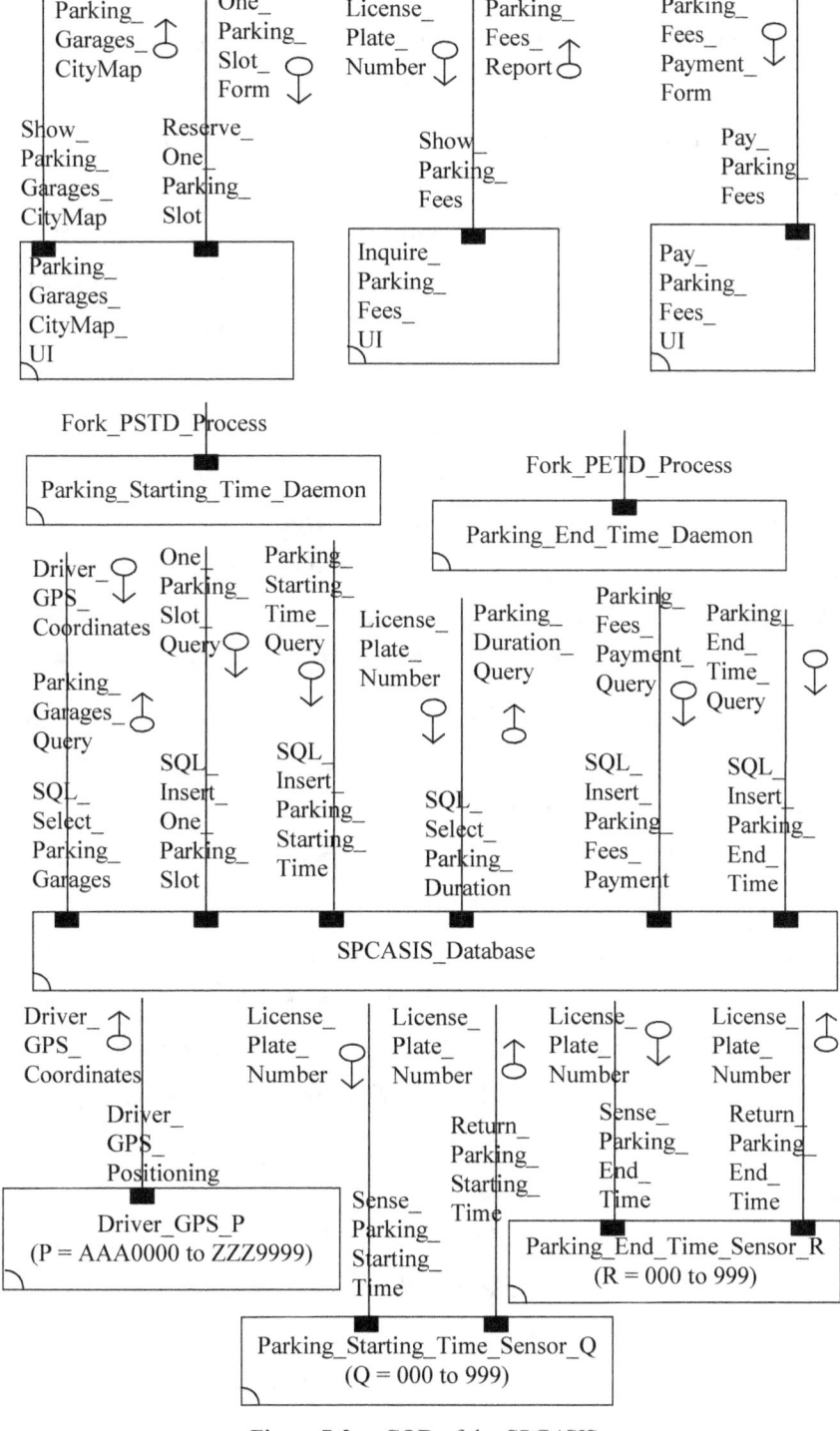

Figure 7-2   COD of the *SPCASIS*

The operation formula of *Show_Parking_Garages_CityMap* is *Show_Parking_Garages_CityMap(Out Parking_Garages_CityMap)*. The operation formula of *Reserve_One_Parking_Slot* is *Reserve_One_Parking_Slot(In One_Parking_Slot_Form)*. The operation formula of *Show_Parking_Fees* is *Show_Parking_Fees(In License_Plate_Number; Out Parking_Fees_Report)*. The operation formula of *Pay_Parking_Fees* is *Pay_Parking_Fees(In Parking_Fees_Payment_Form)*. The operation formula of *Fork_PSTD_Process* is *Fork_PSTD_Process*. The operation formula of *Fork_PETD_Process* is *Fork_PETD_Process*. The operation formula of *SQL_Select_Parking_Garages* is *SQL_Select_Parking_Garages(In Driver_GPS_Coordinates; Out Parking_Garages_Query)*. The operation formula of *SQL_Insert_One_Parking_Slot* is *SQL_Insert_One_Parking_Slot(In One_Parking_Slot_Query)*. The operation formula of *SQL_Insert_Parking_Starting_Time* is *SQL_Insert_Parking_Starting_Time(In Parking_Starting_Time_Query)*. The operation formula of *SQL_Select_Parking_Duration* is *SQL_Select_Parking_Duration(In License_Plate_Number; Out Parking_Duration_Query)*. The operation formula of *SQL_Insert_Parking_Fees_Payment* is *SQL_Insert_Parking_Fees_Payment(In Parking_Fees_Payment_Query)*. The operation formula of *SQL_Insert_Parking_End_Time* is *SQL_Insert_Parking_End_Time(In Parking_End_Time_Query)*. The operation formula of *Driver_GPS_Positioning* is *Driver_GPS_Positioning(Out Driver_GPS_Coordinates)*. The operation formula of *Sense_Parking_Starting_Time* is *Sense_Parking_Starting_Time(In License_Plate_Number)*. The operation formula of *Return_Parking_Starting_Time* is *Return_Parking_Starting_Time(Out License_Plate_Number)*. The operation formula of *Sense_Parking_End_Time* is *Sense_Parking_End_Time(In License_Plate_Number)*. The operation formula of *Return_Parking_End_Time* is *Return_Parking_End_Time(Out License_Plate_Number)*.

Figure 7-3 shows the composite data type specification of the *Parking_Garages_CityMap* output parameter occurring in the *Show_Parking_Garages_CityMap(Out Parking_Garages_CityMap)* operation formula.

Parameter	*Parking_Garages_CityMap*
Data Type	TABLE of Driver_GPS_Coordinates: Text CityMap: Image Parking_Garage: Text Parking_Garage_GPS_Coordinates: Text Available_Slots: Integer End TABLE ;
Instances	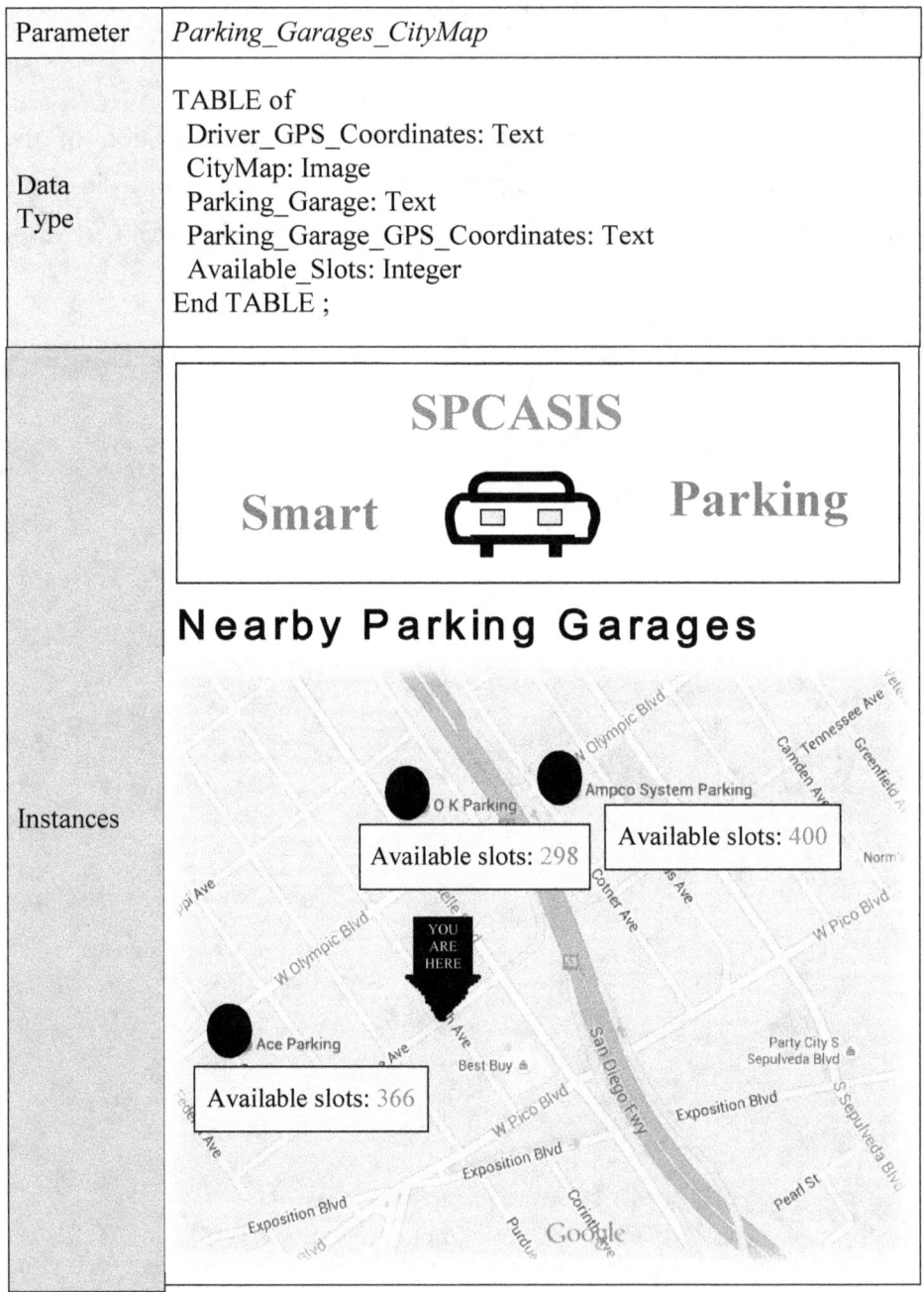

Figure 7-3   Composite Data Type Specification of
Parking_Garages_CityMap

Figure 7-4 shows the composite data type specification of the *One_Parking_Slot_Form* input parameter occurring in the *Reserve_One_Parking_Slot(In One_Parking_Slot_Form)* operation formula.

Parameter	*One_Parking_Slot_Form*
Data Type	TABLE of 　Parking_Garage: Text 　Parking_Slot: Text 　License_Plate_Number: Text 　Reservation_Date_Time: Text End TABLE ;
Instances	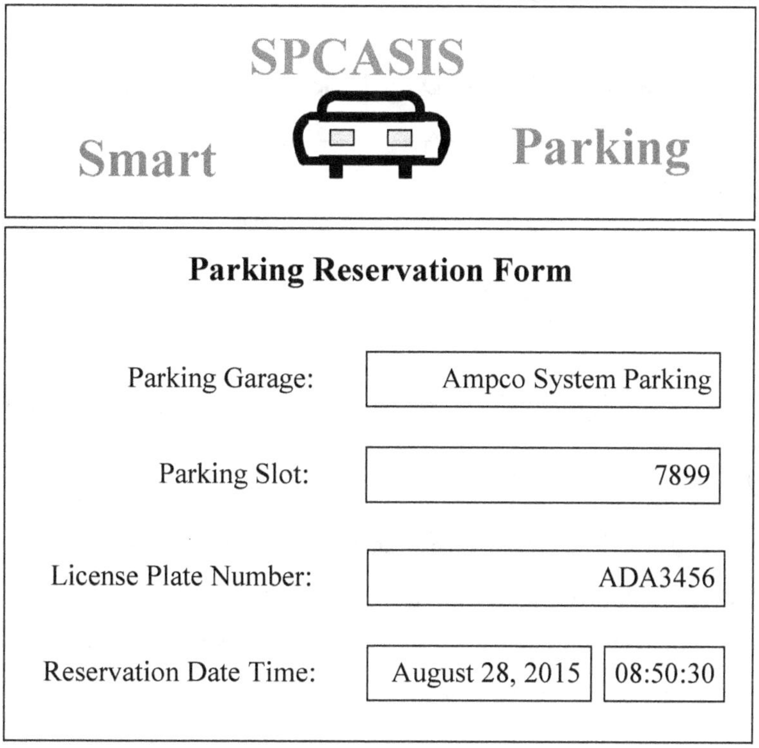

Figure 7-4　Composite Data Type Specification of *One_Parking_Slot_Form*

Figure 7-5 shows the primitive data type specification of the *License_Plate_Number* parameter occurring in the *Show_Parking_Fees(In License_Plate_Number; Out Parking_Fees_Report)*, *SQL_Select_Parking_Duration(In License_Plate_Number; Out Parking_Duration_Query)*, *Sense_Parking_Starting_Time(In License_Plate_Number)*, *Return_Parking_Starting_Time(Out License_Plate_Number)*, *Sense_Parking_End_Time(In License_Plate_Number)* and *Return_Parking_End_Time(Out License_Plate_Number)* operation formulas.

Parameter	Data Type	Instances
License_Plate_Number	Text	ADA3456

Figure 7-5   Primitive Data Type Specification

Figure 7-6 shows the composite data type specification of the *Parking_Fees_Report* output parameter occurring in the *Show_Parking_Fees(In License_Plate_Number; Out Parking_Fees_Report)* operation formula.

Parameter	*Parking_Fees_Report*
Data Type	TABLE of Parking_Garage: Text License_Plate_Number: Text Parking_Duration: Integer Total Fees: Real End TABLE ;
Instances	SPCASIS Smart Parking **Parking Fees** Parking Garage: Ampco System Parking License Plate Number: ADA3456 Parking Duration (Minutes): 755 Total Fees: $20.70

Figure 7-6 Composite Data Type Specification of *Parking_Fees_Report*

Figure 7-7 shows the composite data type specification of the *Parking_Fees_Payment_Form* input parameter occurring in the *Pay_Parking_Fees(In Parking_Fees_Payment_Form)* operation formula.

Parameter	*Parking_Fees_Payment_Form*
Data Type	TABLE of Parking_Garage: Text Date_Time: Text License_Plate_Number: Text Amount: Real End TABLE ;
Instances	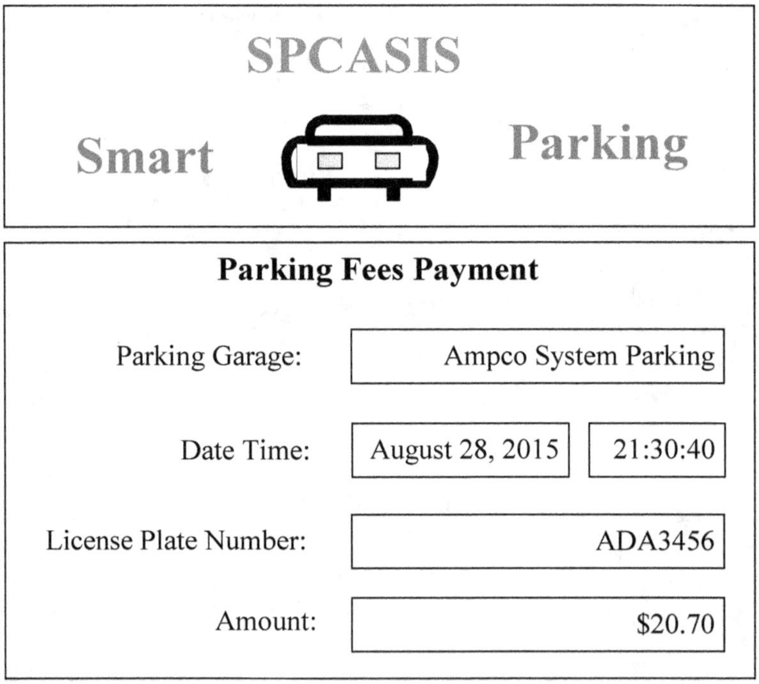

Figure 7-7   Composite Data Type Specification of
Parking_Fees_Payment_Form

Figure 7-8 shows the primitive data type specification of the *Driver_GPS_Coordinates* parameter occurring in the *SQL_Select_Parking_Garages(In Driver_GPS_Coordinates; Out Parking_Garages_Query)* and *Driver_GPS_Positioning(Out Driver_GPS_Coordinates)* operation formulas.

Parameter	Data Type	Instances
Driver_ GPS_ Coordinates	Text	34.036964, -118.441646

Figure 7-8   Primitive Data Type Specification

Figure 7-9 shows the composite data type specification of the *Parking_Garages_Query* output parameter occurring in the *SQL_Select_Parking_Garages(In Driver_GPS_Coordinates; Out Parking_Garages_Query)* operation formula.

Parameter	*Parking_Garages_Query*

Data Type	TABLE of
	  Parking_Garage: Text
	  Parking_Garage_GPS_Coordinates: Text
	  Available_Slots: Integer
	End TABLE ;

Instances	

Parking_ Garage	Parking_ Garage_ GPS_ Coordinates	Available_ Slots
Ampco System Parking	34.039311, -118.438579	400
O K Parking	34.03919, -118.441495	298
Ace Parking	34.035844 , -118.444571	366

Figure 7-9   Composite Data Type Specification of *Parking_Garages_Query*

Figure 7-10 shows the composite data type specification of the *One_Parking_Slot_Query* input parameter occurring in the *SQL_Insert_One_Parking_Slot(In One_Parking_Slot_Query)* operation formula.

Parameter	*One_Parking_Slot_Query*

Data Type	TABLE of
	  Parking_Garage: Text
	  Parking_Slot: Text
	  License_Plate_Number: Text
	  Reservation_Date_Time: Text
	End TABLE ;

Instances	<table><tr><th>Parking_ Garage</th><th>Parking_ Slot</th><th>License_ Plate_ Number</th><th>Reservation_ Date_ Time</th></tr><tr><td>Ampco System Parking</td><td>7899</td><td>ADA3456</td><td>20150828, 08:50:30</td></tr></table>

Figure 7-10   Composite Data Type Specification of *One_Parking_Slot_Query*

Figure 7-11 shows the composite data type specification of the *Parking_Starting_Time_Query* input parameter occurring in the *SQL_Insert_Parking_Starting_Time(In   Parking_Starting_Time_Query)* operation formula.

Parameter	*Parking_Starting_Time_Query*			

Data Type	TABLE of
	  Parking_Garage: Text
	  Parking_Slot: Text
	  License_Plate_Number: Text
	  Parking_Starting_Date_Time: Text
	End TABLE ;			

Instances	Parking_ Garage	Parking_ Slot	License_ Plate_ Number	Parking_ Starting_ Date_ Time
	Ampco System Parking	7899	ADA3456	20150828, 08:55:40

Figure 7-11   Composite Data Type Specification of
Parking_Starting_Time_Query

Figure 7-12 shows the composite data type specification of the *Parking_Duration_Query* output parameter occurring in the *SQL_Select_Parking_Duration(In License_Plate_Number; Out Parking_Duration_Query)* operation formula.

Parameter	*Parking_Duration_Query*
Data Type	TABLE of 　Parking_Garage: Text 　License_Plate_Number: Text 　Parking_Duration: Integer End TABLE ;
Instances	

Parking_ Garage	License_ Plate_ Number	Parking_ Duration (Minutes)
Ampco System Parking	ADA3456	755

Figure 7-12　Composite Data Type Specification of
Parking_Duration_Query

Figure 7-13 shows the composite data type specification of the *Parking_Fees_Payment_Query* input parameters occurring in the *SQL_Insert_Parking_Fees_Payment(In Parking_Fees_Payment_Query)* operation formula.

Parameter	*Parking_Fees_Payment_Query*

Data Type	TABLE of
	  Parking_Garage: Text
	  Date_Time: Text
	  License_Plate_Number: Text
	  Amount: Real
	End TABLE ;

Instances	

Parking_ Garage	Date_ Time	License_ Plate_ Number	Amount
Ampco System Parking	20150828, 21:30:40	ADA3456	20.70

Figure 7-13   Composite Data Type Specification of
Parking_Fees_Payment_Query

Figure 7-14 shows the composite data type specification of the *Parking_End_Time_Query* input parameters occurring in the *SQL_Insert_Parking_End_Time(In Parking_End_Time_Query)* operation formula.

Parameter	*Parking_End_Time_Query*			
Data Type	TABLE of Parking_Garage: Text Parking_Slot: Text License_Plate_Number: Text Parking_End_Date_Time: Text End TABLE ;			
Instances	Parking_ Garage	Parking_ Slot	License_ Plate_ Number	Parking_ End_ Date_ Time
	Ampco System Parking	7899	ADA3456	20150828, 21:36:50

Figure 7-14   Composite Data Type Specification of
Parking_End_Time_Query

7-3 Interaction Flow Diagram

The overall behavior of the *Smart Parking Cloud Applications and Services IoT System* (SPCASIS) includes five individual behaviors: *Finding_and_Reserving_a_Vacant_Parking_Slot*, *Sensing_Parking_Starting_Time*, *Inquiring_Parking_Fees*, *Paying_Parking_Fees*, *Sensing_Parking_End_Time*. Each individual behavior is represented by an execution path. SyRS 2.0 uses an IFD to specify each one of these execution paths.

Figure 7-15 shows an IFD of the *Finding_and_Reserving_a_Nearby_Vacant_Parking_Slot* behavior. First, actor *Driver* interacts with the *Parking_Garages_CityMap_UI* component through the *Show_Parking_Garages_CityMap* operation call interaction. Next, component *Parking_Garages_CityMap_UI* interacts with the *Driver_GPS_P (P = AAA0000 to ZZZ9999)* component through the *Driver_GPS_Positioning* operation call interaction, carrying the *Driver_GPS_Coordinates* output parameter. Continuingly, component *Parking_Garages_CityMap_UI* interacts with the *SPCASIS_Database* component through the *SQL_Select_Parking_Garages* operation call interaction, carrying the *Driver_GPS_Coordinates* input parameter and *Parking_Garages_Query* output parameter. Continuingly, actor *Driver* interacts with the *Parking_Garages_CityMap_UI* component through the *Show_Parking_Garages_CityMap* operation return interaction, carrying the *Parking_Garages_CityMap* output parameter. Continuingly, actor *Driver* interacts with the *Parking_Garages_CityMap_UI* component through the *Reserve_One_Parking_Slot* operation call interaction, carrying the *One_Parking_Slot_Form* input parameter. Finally, component *Parking_Garages_CityMap_UI* interacts with the *SPCASIS_Database* component through the *SQL_Insert_One_Parking_Slot* operation call interaction, carrying the *One_Parking_Slot_Query* input parameter.

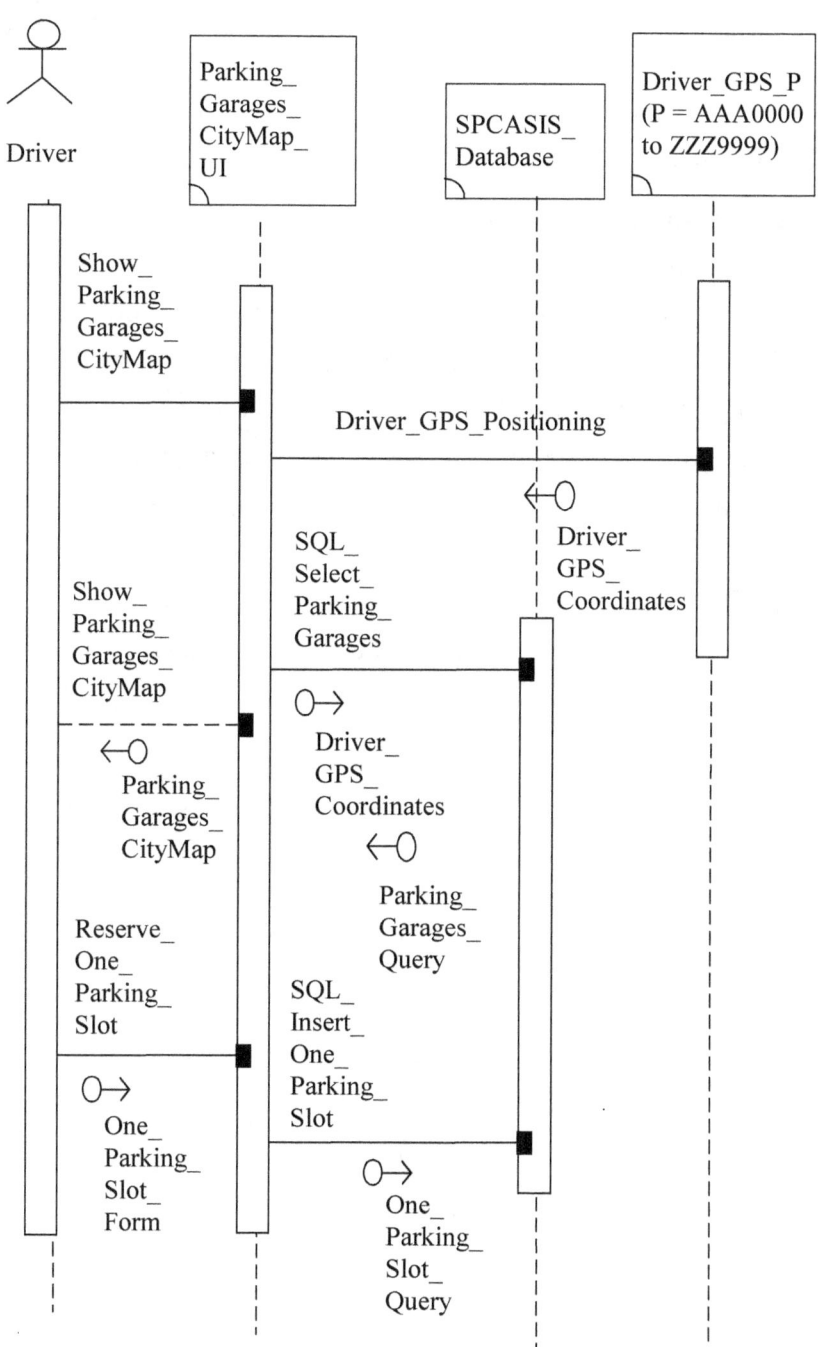

Figure 7-15   IFD of the
Finding_and_Reserving_a_Vacant_Parking_Slot Behavior

Figure 7-16 shows an IFD of the *Sensing_Parking_Starting_Time* behavior. First, actor *Server_Root* interacts with the *Parking_Starting_Time_Daemon* component through the *Fork_PSTD_Process* operation call interaction. Next, actor *Car_Tag* interacts with the *Parking_Starting_Time_Sensor_Q (Q = 000 to 999)* component through the *Sense_Parking_Starting_Time* operation call interaction, carrying the *License_Plate_Number* input parameter. Continuingly, component *Parking_Starting_Time_Daemon* interacts with the *Parking_Starting_Time_Sensor_Q (Q = 000 to 999)* component through the *Return_Parking_Starting_Time* operation call interaction, carrying the *License_Plate_Number* output parameter. Finally, component *Parking_Starting_Time_Daemon* interacts with the *SPCASIS_Database* component through the *SQL_Insert_Parking_Starting_Time* operation call interaction, carrying the *Parking_Starting_Time_Query* input parameter.

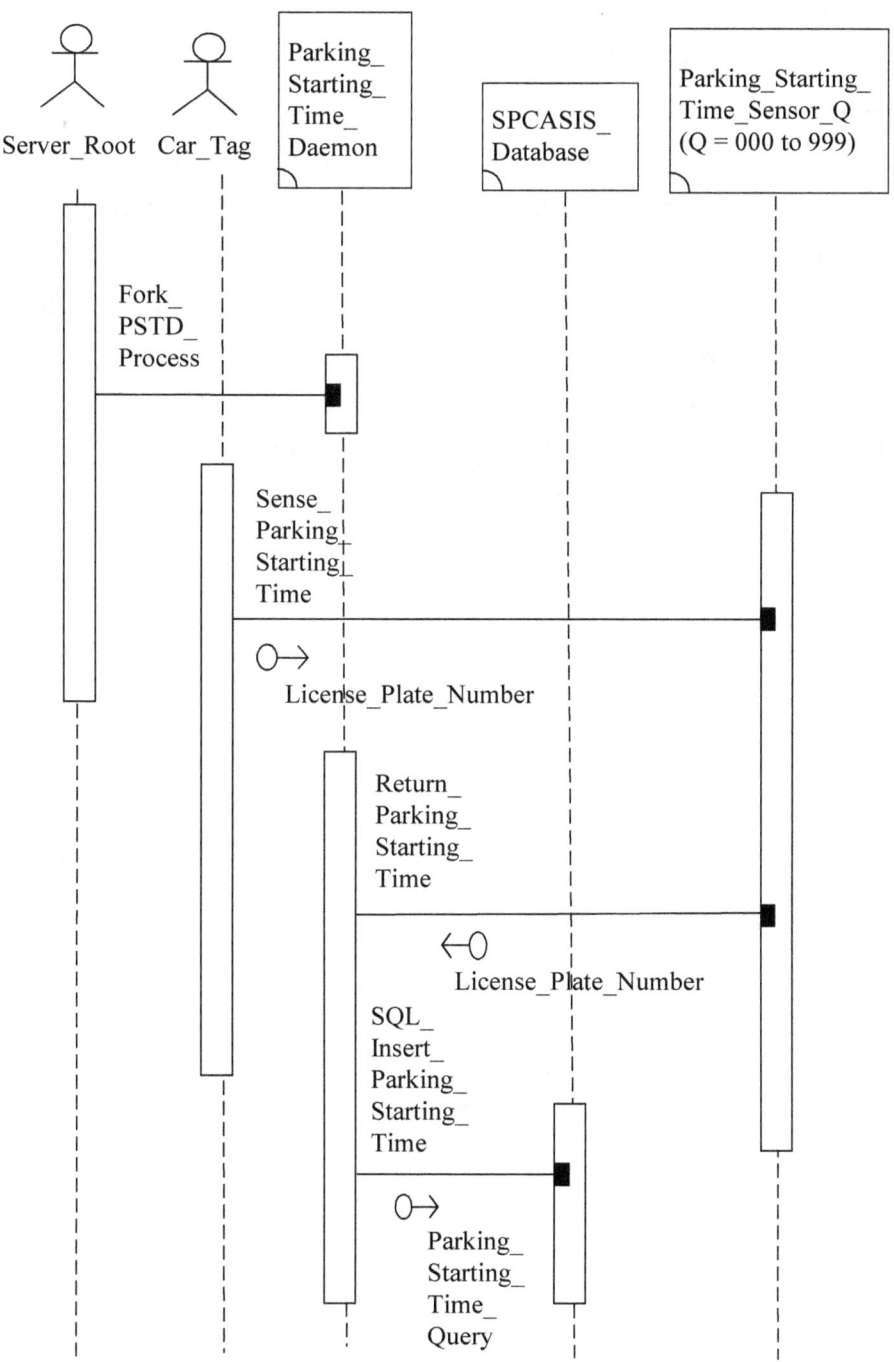

Figure 7-16   IFD of the *Sensing_Parking_Starting_Time* Behavior

Figure 7-17 shows an IFD of the *Inquiring_Parking_Fees* behavior. First, actor *Driver* interacts with the *Inquire_Parking_Fees_UI* component through the *Show_Parking_Fees* operation call interaction, carrying the *License_Plate_Number* input parameter. Next, component *Inquire_Parking_Fees_UI* interacts with the *SPCASIS_Database* component through the *SQL_Select_Parking_Duration* operation call interaction, carrying the *License_Plate_Number* input parameter and the *Parking_Duration_Query* output parameter. Finally, actor *Driver* interacts with the *Inquire_Parking_Fees_UI* component through the *Show_Parking_Fees* operation return interaction, carrying the *Parking_Fees_Report out*put parameter.

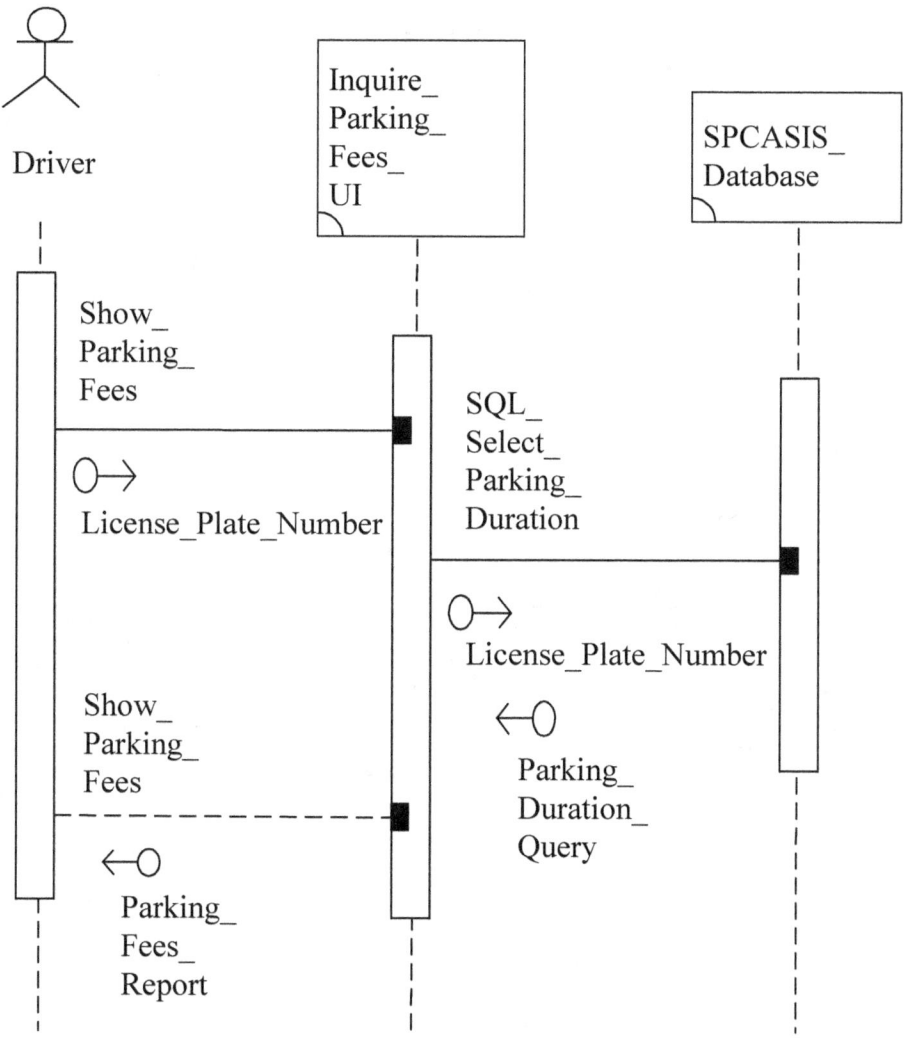

Figure 7-17   IFD of the *Inquiring_Parking_Fees* Behavior

Figure 7-18 shows an IFD of the *Paying_Parking_Fees* behavior. First, actor *Driver* interacts with the *Pay_Parking_Fees_UI* component through the *Pay_Parking_Fees* operation call interaction, carrying the

Parking_Fees_Payment_Form input parameters. Finally, component *Pay_Parking_Fees_UI* interacts with the *SPCASIS_Database* component through the *SQL_Insert_Parking_Fees_Payment* operation call interaction, carrying the *Parking_Fees_Payment_Query* input parameter.

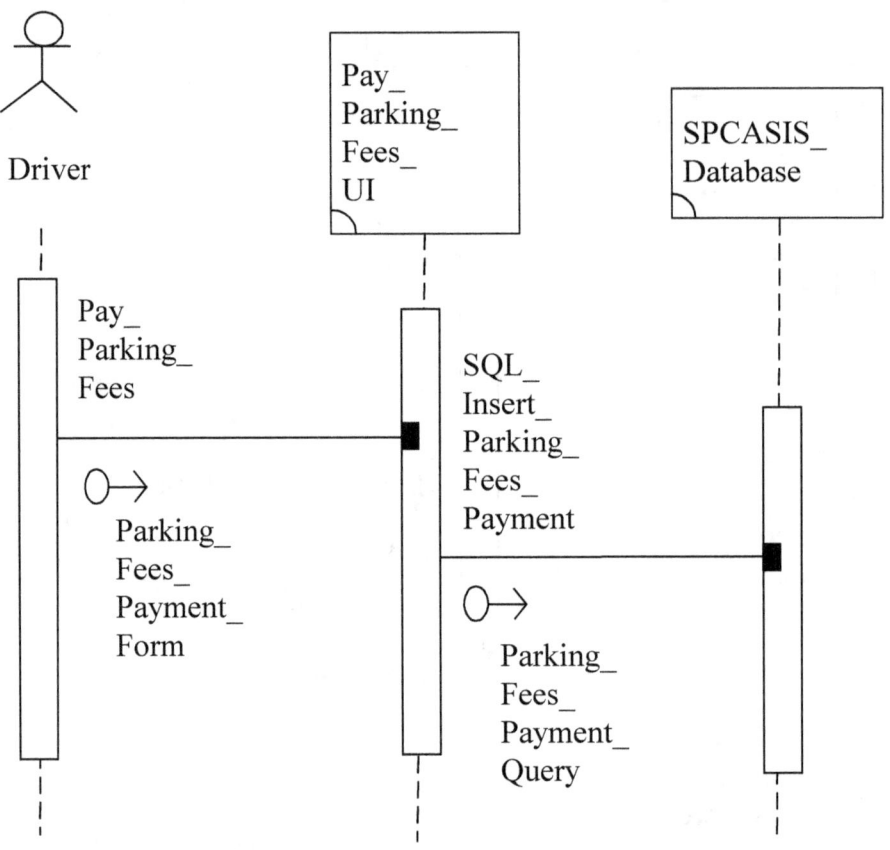

Figure 7-18   IFD of the *Paying_Parking_Fees* Behavior

Figure 7-19 shows an IFD of the *Sensing_Parking_End_Time* behavior. First, actor *Server_Root* interacts with the *Parking_End_Time_Daemon* component through the *Fork_PETD_Process* operation call interaction. Next, actor *Car_Tag* interacts with the *Parking_End_Time_Sensor_R (R = 000 to 999)* component through the *Sense_Parking_End_Time* operation call interaction, carrying the *License_Plate_Number* input parameter. Continuingly, component *Parking_End_Time_Daemon* interacts with the *Parking_End_Time_Sensor_R (R = 000 to 999)* component through the *Return_Parking_End_Time* operation call interaction, carrying the *License_Plate_Number* output parameter. Finally, component *Parking_End_Time_Daemon* interacts with the *SPCASIS_Database* component through the *SQL_Insert_Parking_End_Time* operation call interaction, carrying the *Parking_End_Time_Query* input parameter.

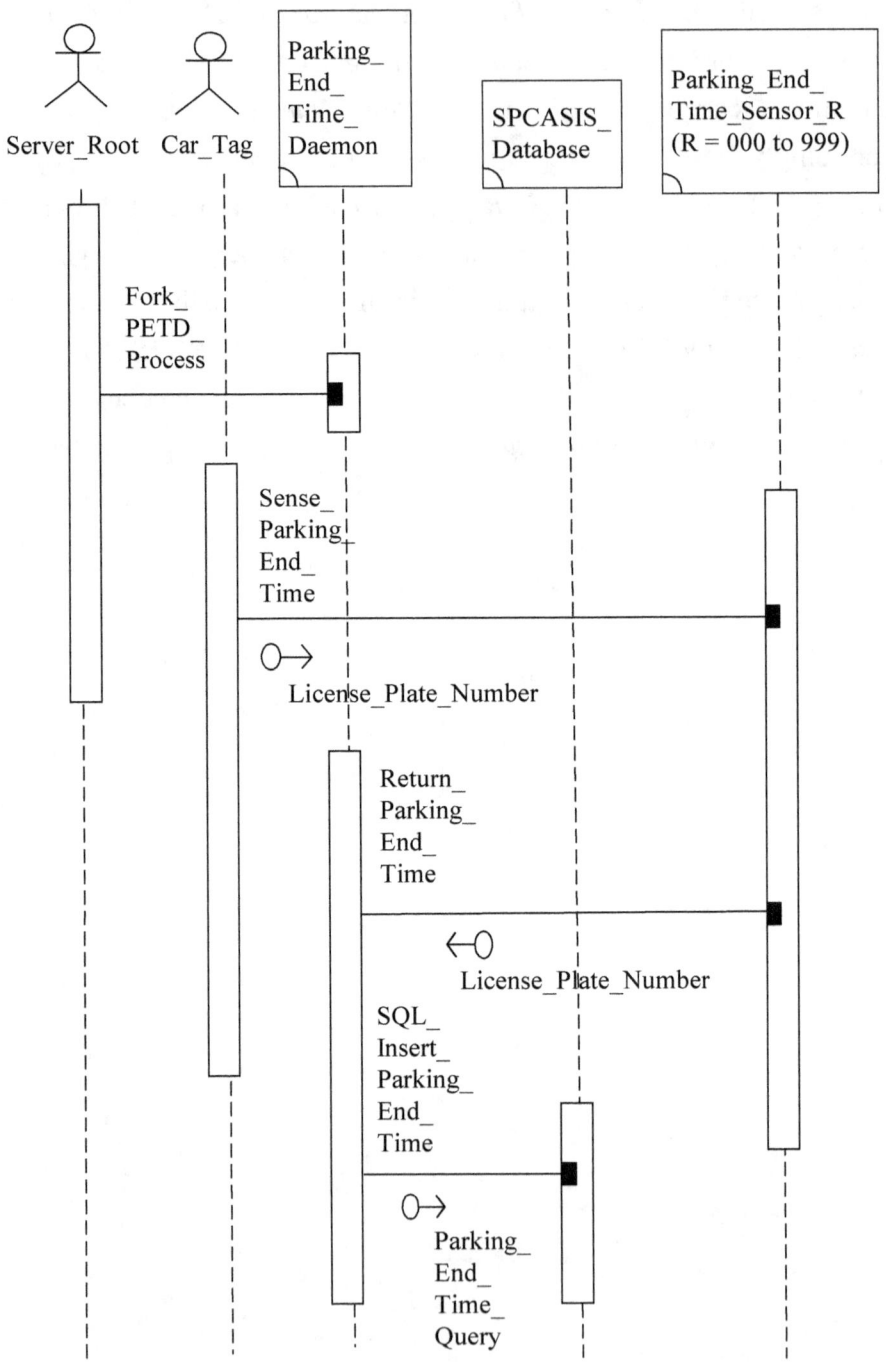

Figure 7-19   IFD of the *Sensing_Parking_End_Time* Behavior

Chapter 8: System Requirement Specification 2.0 of the Landslide Prevention and Relief Cloud Applications and Services IoT System

A landslide occurs when part of a natural slope is unable to support its own weight [Coro08, Fell08]. For instance, wet soil material on a slippery surface underneath, can become heavy with rainwater and slide down due to its increased weight. A landslide is a downward or outward movement of soil, rock or vegetation, under the influence of gravity. This movement can occur in many ways. It can be a fall, topple, slide, spread or flow. Velocity of the movement may range from very slow to fast. The mass of moving material can destroy property along its path of movement and cause death to human beings and livestock.

Advancements in cloud applications and services systems present enormous potential for accurate monitoring and providing service to the customers and administrators. A cost effective solution to this service can be provided by wireless sensor networks which consists of larger number of sensors placed on the existing land areas without installing expensive cabling and are capable of adjusting with the easily available sensors. The information obtained from each sensor is processed collaboratively to evaluate meaning metrics such as landslide signs of all land areas. Landslide prevention and relief cloud applications and services IoT system (LPRCASIS) covers a great deal of land areas. A land area consists of a group of landslide signs sensors.

Behaviors of LPRCASIS consist of: a) *Sensing_Landslide_Signs* behavior, b) *Recording_Occurring_Landslide* behavior, c) *Alerts_Notifying* behavior, d) *Recording_Emergency_Response_Starting_Time* behavior and e) *Recording_Emergency_Response_End_Time* behavior.

Using the structure-behavior coalescence (SBC) approach, we shall go through: a) architecture hierarchy diagram, b) component operation diagram and c) interaction flow diagram, to accomplish the system requirements specification (SyRS) 2.0 for LPRCASIS.

8-1 Architecture Hierarchy Diagram

SyRS 2.0 uses an architecture hierarchy diagram (AHD) to specify the multi-level composition and decomposition of the *Landslide Prevention and Relief Cloud Applications and Services IoT System* (LPRCASIS) as shown in Figure 8-1. In the figure, *LPRCASIS* is composed of *Application_Layer*, *Data_Layer* and *Technology_Layer*; *Application_Layer* is composed of *Presentation_Layer* and *Logic_Layer*; *Presentation_Layer* is composed of *Occurring_Landslide_UI*, *Alerts_Notifying_UI*, *Emergency_Response_Starting_Time_UI* and *Emergency_Response_End_Time_UI*; *Logic_Layer* is composed of *Landslide_Signs_Daemon*; *Data_Layer* is composed of *LPRCASIS_Database*; *Technology_Layer* is composed of *Landslide_Signs_Sensor_N (N = A0000 to Z9999)*.

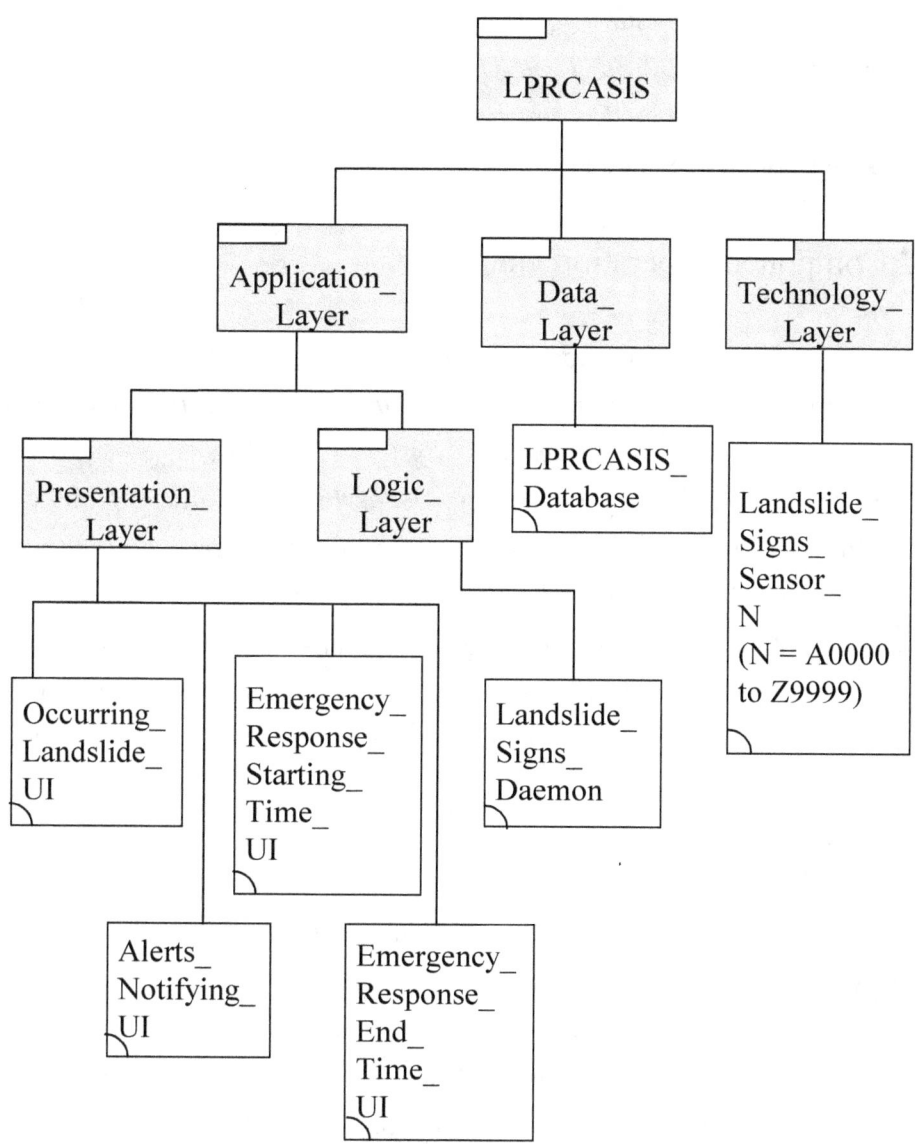

Figure 8-1   AHD of the *LPRCASIS*

In Figure 8-1, *LPRCASIS, Application_Layer, Presentation_Layer, Logic_Layer, Data_Layer* and *Technology_Layer* are aggregated systems while   *Occurring_Landslide_UI,*   *Alerts_Notifying_UI,*

Emergency_Response_Starting_Time_UI,
Emergency_Response_End_Time_UI, *Landslide_Signs_Daemon*,
LPRCASIS_Database and *Landslide_Signs_Sensor_N (N = A0000 to Z9999)* are non-aggregated systems.

8-2 Component Operation Diagram

SyRS 2.0 uses a component operation diagram (COD) to specify the operations of all components of the *Landslide Prevention and Relief Cloud Applications and Services IoT System* (LPRCASIS) as shown in Figure 8-2. In the figure, component *Occurring_Landslide_UI* has one operation: *Input_Occurring_Landslide*; component *Alerts_Notifying_UI* has two operations: *Show_All_Alerts* and *Display_Alerts*; component *Emergency_Response_Starting_Time_UI* has one operation: *Input_Emergency_Response_Starting_Time*; component *Emergency_Response_End_Time_UI* has one operation: *Input_Emergency_Response_End_Time*; component *Landslide_Signs_Daemon* has one operation: *Fork_LSD_Process*; component *LPRCASIS_Database* has five operations: *SQL_Insert_Landslide_Signs*, *SQL_Insert_Occurring_Landslide*, *SQL_Select_Landslides_Signs_for_Alerts_Analysis*, *SQL_Insert_Emergency_Response_Starting_Time* and *SQL_Insert_Emergency_Response_End_Time*; component *Landslide_Signs_Sensor_N (N = A0000 to Z9999)* has two operations: *Sense_Landslide_Signs* and *Return_Landslide_Signs*.

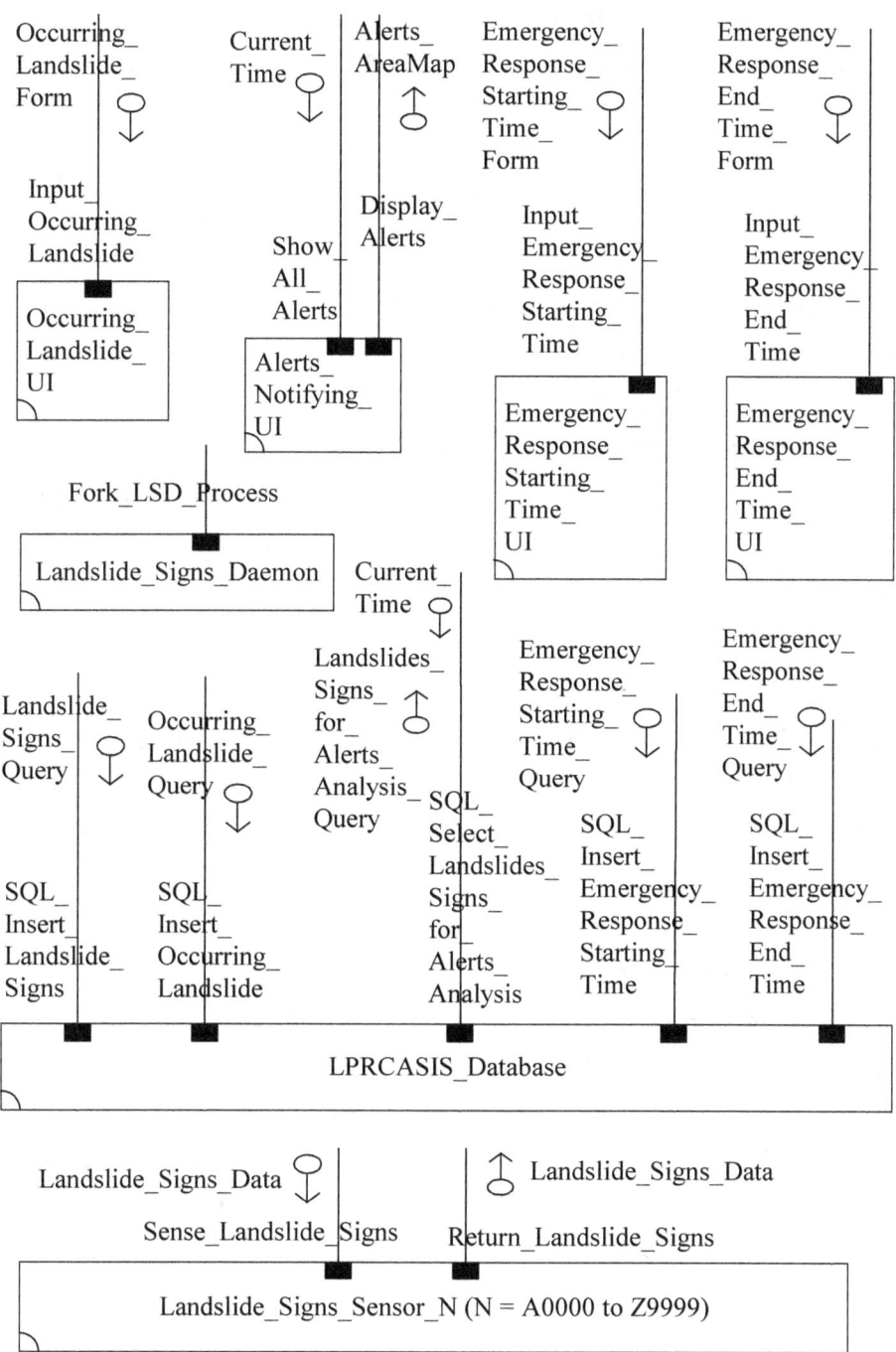

Figure 8-2   COD of the *LPRCASIS*

The operation formula of *Input_Occurring_Landslide* is *Input_Occurring_Landslide(In Occurring_Landslide_Form)*. The operation formula of *Show_All_Alerts* is *Show_All_Alerts(In Current_Time)*. The operation formula of *Display_Alerts* is *Display_Alerts(Out Alerts_AreaMap)*. The operation formula of *Input_Emergency_Response_Starting_Time* is *Input_Emergency_Response_Starting_Time(In Emergency_Response_Starting_Time_Form)*. The operation formula of *Input_Emergency_Response_End_Time* is *Input_Emergency_Response_End_Time(In Emergency_Response_End_Time_Form)*. The operation formula of *Fork_LSD_Process* is *Fork_LSD_Process*. The operation formula of *SQL_Insert_Landslide_Signs* is *SQL_Insert_Landslide_Signs(In Landslide_Signs_Query)*. The operation formula of *SQL_Insert_Occurring_Landslide* is *SQL_Insert_Occurring_Landslide(In Occurring_Landslide_Query)*. The operation formula of *SQL_Select_Landslides_Signs_for_Alerts_Analysis* is *SQL_Select_Landslides_Signs_for_Alerts_Analysis(In Current_Time; Out Landslides_Signs_for_Alerts_Analysis_Query)*. The operation formula of *SQL_Insert_Emergency_Response_Starting_Time* is *SQL_Insert_Emergency_Response_Starting_Time(In Emergency_Response_Starting_Time_Query)*. The operation formula of *SQL_Insert_Emergency_Response_End_Time* is *SQL_Insert_Emergency_Response_End_Time(In Emergency_Response_End_Time_Query)*. The operation formula of *Sense_Landslide_Signs* is *Sense_Landslide_Signs(In Landslide_Signs_Data)*. The operation formula of *Return_Landslide_Signs* is *Return_Landslide_Signs(Out Landslide_Signs_Data)*.

Figure 8-3 shows the composite data type specification of the

Occurring_Landslide_Form input parameter occurring in the *Input_Occurring_Landslide(In Occurring_Landslide_Form)* operation formula.

Parameter	*Occurring_Landslide_Form*
Data Type	TABLE of Occurring_Time: Text Landslides_Signs_GPS_Coordinates: Text Landslide_Signs_Data: Integer End TABLE ;
Instances	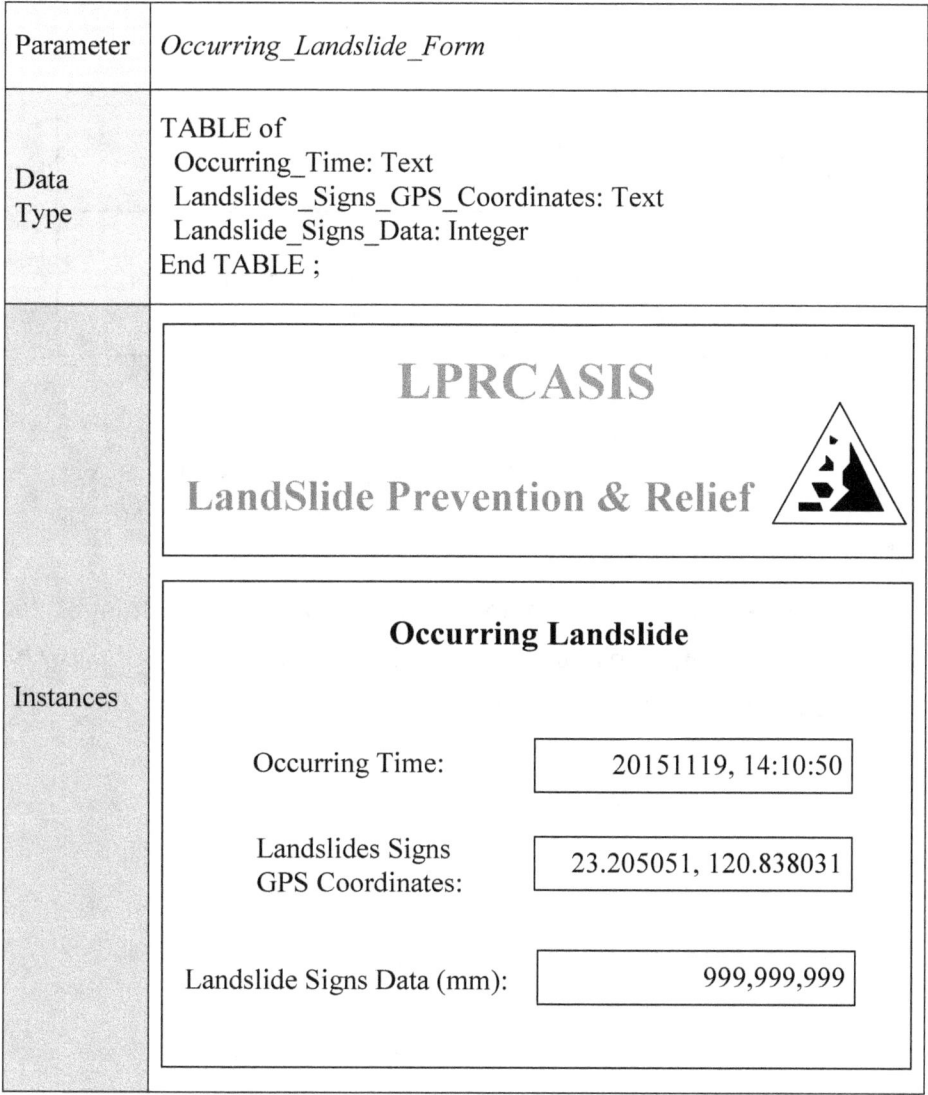

Figure 8-3   Composite Data Type Specification of
Occurring_Landslide_Form

Figure 8-4 shows the primitive data type specification of the

Current_Time parameter occurring in the *Show_All_Alerts(In Current_Time)* and *SQL_Select_Landslides_Signs_for_Alerts_Analysis(In Current_Time; Out Landslides_Signs_for_Alerts_Analysis_Query)* operation formulas.

Parameter	Data Type	Instances
Current_ Time	Text	20151129, 14:11:00

Figure 8-4   Primitive Data Type Specification

Figure 8-5 shows the composite data type specification of the *Alerts_AreaMap* output parameter occurring in the *Display_Alerts(Out Alerts_AreaMap)* operation formula.

Parameter	*Alerts_AreaMap*

Data Type	TABLE of
	  AreaMap: Image
	  Landslides_Signs_GPS_Coordinates: Text
	  Emergency_Response: Boolean
	  Landslide_Signs_Data: Integer
	End TABLE ;

Instances	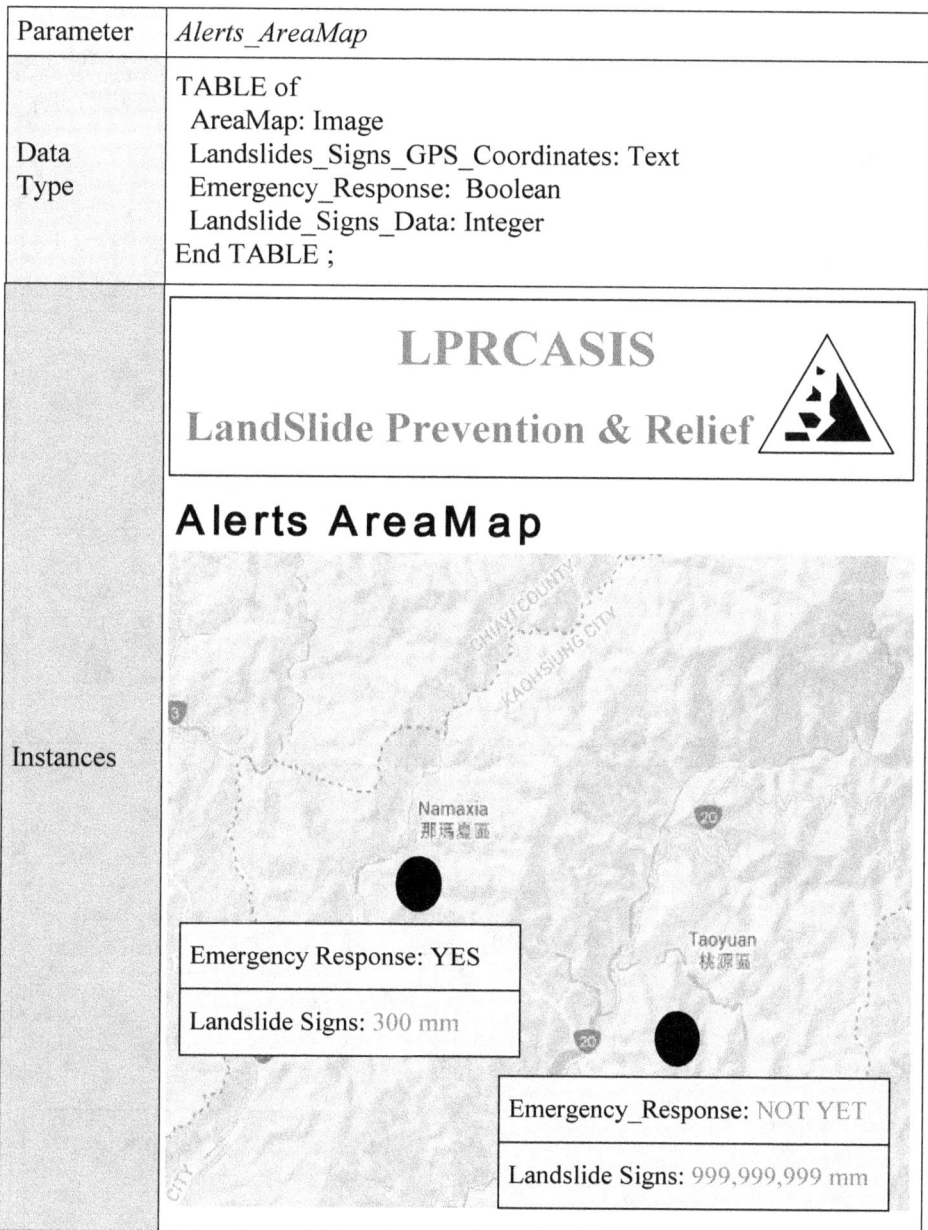

Figure 8-5   Composite Data Type Specification of *Alerts_AreaMap*

Figure 8-6 shows the composite data type specification of the *Emergency_Response_Starting_Time_Form* input parameter occurring in the *Input_Emergency_Response_Starting_Time(In Emergency_Response_Starting_Time_Form)* operation formula.

Parameter	*Emergency_Response_Starting_Time_Form*

Data Type	TABLE of
	  Landslide_GPS_Coordinates: Text
	  Emergency_Response_Starting_Time: Text
	  Actions: Text
	End TABLE ;

Instances	**LPRCASIS** **LandSlide Prevention & Relief** **Emergency Response Starting Time** Landslide_GPS_Coordinates: 23.254267, 120.718555 Emergency_Response_Starting_Time: 20151119, 14:12:00 Actions: Inform affected neighbours; Begin evacuation; Remove debris

Figure 8-6   Composite Data Type Specification of
Emergency_Response_Starting_Time_Form

Figure 8-7 shows the composite data type specification of the *Emergency_Response_End_Time_Form* input parameter occurring in the *Input_Emergency_Response_End_Time(In Emergency_Response_End_Time_Form)* operation formula.

Parameter	*Emergency_Response_End_Time_Form*
Data Type	TABLE of Landslide_GPS_Coordinates: Text Emergency_Response_End_Time: Text End TABLE ;
Instances	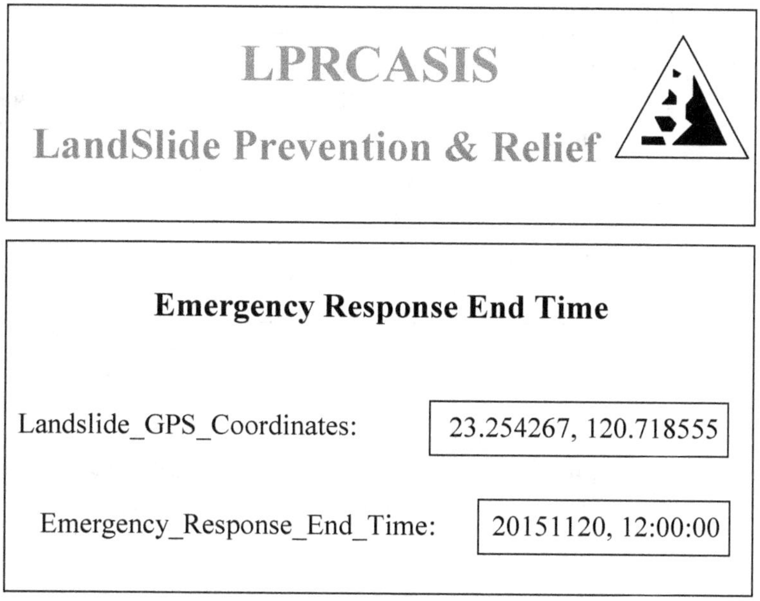

Figure 8-7   Composite Data Type Specification of
Emergency_Response_End_Time_Form

Figure 8-8 shows the composite data type specification of the *Landslide_Signs_Query* input parameter occurring in the *SQL_Insert_Landslide_Signs(In Landslide_Signs_Query)* operation formula.

Parameter	*Landslide_Signs_Query*		
Data Type	TABLE of Sensing_Time: Text Landslide_Signs_GPS_Coordinates: Text Landslide_Signs_Data: Integer End TABLE ;		
Instances	Sensing_ Time	Landslide_ Signs_ GPS_ Coordinates	Landslide_ Signs_ Data (mm)
	20151119, 14:10:20	23.254267, 120.718555	300

Figure 8-8   Composite Data Type Specification of *Landslide_Signs_Query*

Figure 8-9 shows the composite data type specification of the *Occurring_Landslide_Query* input parameter occurring in the *SQL_Insert_Occurring_Landslide(In Occurring_Landslide_Query)* operation formula.

Parameter	*Occurring_Landslide_Query*		

Data Type	TABLE of
	  Occurring_Time: Text
	  Landslides_Signs_GPS_Coordinates: Text
	  Landslide_Signs_Data: Integer
	End TABLE ;		

Instances	Occurring_ Time	Landslides_ Signs_ GPS_ Coordinates	Landslide_ Signs_ Data (mm)
	20151119, 14:10:50	23.205051, 120.838031	999,999,999

Figure 8-9   Composite Data Type Specification of
Occurring_Landslide_Query

Figure 8-10 shows the composite data type specification of the *Landslides_Signs_for_Alerts_Analysis_Query* output parameter occurring in the *SQL_Select_Landslides_Signs_for_Alerts_Analysis(In Current_Time; Out Landslides_Signs_for_Alerts_Analysis_Query)* operation formula.

Parameter	*Landslides_Signs_for_Alerts_Analysis_Query*

Data Type	TABLE of
	  Landslides_Signs_GPS_Coordinates: Text
	  Sensing_Or_Occurring_Time: Text
	  Emergency_Response_Starting_Time: Text
	  Landslide_Signs_Data: Integer
	End TABLE ;

Instances	Landslides_ Signs_ GPS_ Coordinates	Sensing_ Or_ Occurring_ Time	Emergency_ Response_ Starting_ Time	Landslide_ Signs_ Data (mm)
	23.254267, 120.718555	20151119, 14:10:20	20151119, 14:12:00	200
	23.205051, 120.838031	20151119, 14:10:50	NULL	999,999,999

Figure 8-10   Composite Data Type Specification of
Landslides_Signs_for_Alerts_Analysis_Query

Figure 8-11 shows the composite data type specification of the *Emergency_Response_Starting_Time_Query* input parameter occurring in the *SQL_Insert_Emergency_Response_Starting_Time(In Emergency_Response_Starting_Time_Query)* operation formula.

Parameter	*Emergency_Response_Starting_Time_Query*
Data Type	TABLE of Landslide_GPS_Coordinates: Text Emergency_Response_Starting_Time: Text Actions: Text End TABLE ;

Instances			
	Landslide_ GPS_ Coordinates	Emergency_ Response_ Starting_ Time	Actions
	23.254267, 120.718555	20151119, 14:12:00	Inform affected neighbours; Begin evacuation; Remove debris

Figure 8-11   Composite Data Type Specification of
Emergency_Response_Starting_Time_Query

Figure 8-12 shows the composite data type specification of the
Emergency_Response_End_Time_Query input parameter occurring in the
SQL_Insert_Emergency_Response_End_Time(In
Emergency_Response_End_Time_Query) operation formula.

126

Parameter	*Emergency_Response_End_Time_Query*

Data Type	TABLE of
	  Landslide_GPS_Coordinates: Text
	  Emergency_Response_End_Time: Text
	End TABLE ;

Instances	<table><tr><td>Landslide_ GPS_ Coordinates</td><td>Emergency_ Response_ End_ Time</td></tr><tr><td>23.254267, 120.718555</td><td>20151120, 13:00:00</td></tr></table>

Figure 8-12   Composite Data Type Specification of
Emergency_Response_End_Time_Query

Figure 8-13 shows the primitive data type specification of the *Landslide_Signs_Data* parameter occurring in the *Sense_Landslide_Signs(In Landslide_Signs_Data)* and *Return_Landslide_Signs(Out Landslide_Signs_Data)* operation formulas.

Parameter	Data Type	Instances
Landslide_ Signs_ Data (mm)	Text	300

Figure 8-13   Primitive Data Type Specification

8-3 Interaction Flow Diagram

The overall behavior of the *Landslide Prevention and Relief Cloud Applications and Services IoT System* (LPRCASIS) includes five individual behaviors: *Sensing_Landslide_Signs*, *Recording_Occurring_Landslide*, *Alerts_Notifying*, *Recording_Emergency_Response_Starting_Time*, *Recording_Emergency_Response_End_Time*. Each individual behavior is represented by an execution path. SyRS 2.0 uses an IFD to specify each one of these execution paths.

Figure 8-14 shows an IFD of the *Sensing_Landslide_Signs* behavior. First, actor *Server_Root* interacts with the *Landslide_Signs_Daemon* component through the *Fork_LSD_Process* operation call interaction. Next, actor *Landslide_Signs* interacts with the *Landslide_Signs_Sensor_N (N = A0000 to Z9999)* component through the *Sense_Landslide_Signs* operation call interaction, carrying the

Landslide_Signs_Data input parameter. Continuingly, component *Landslide_Signs_Daemon* interacts with the *Landslide_Signs_Sensor_N (N = A0000 to Z9999)* component through the *Return_Landslide_Signs* operation call interaction, carrying the *Landslide_Signs_Data* output parameter. Finally, component *Landslide_Signs_Daemon* interacts with the *LPRCASIS_Database* component through the *SQL_Insert_Landslide_Signs* operation call interaction, carrying the *Landslide_Signs_Query* input parameter.

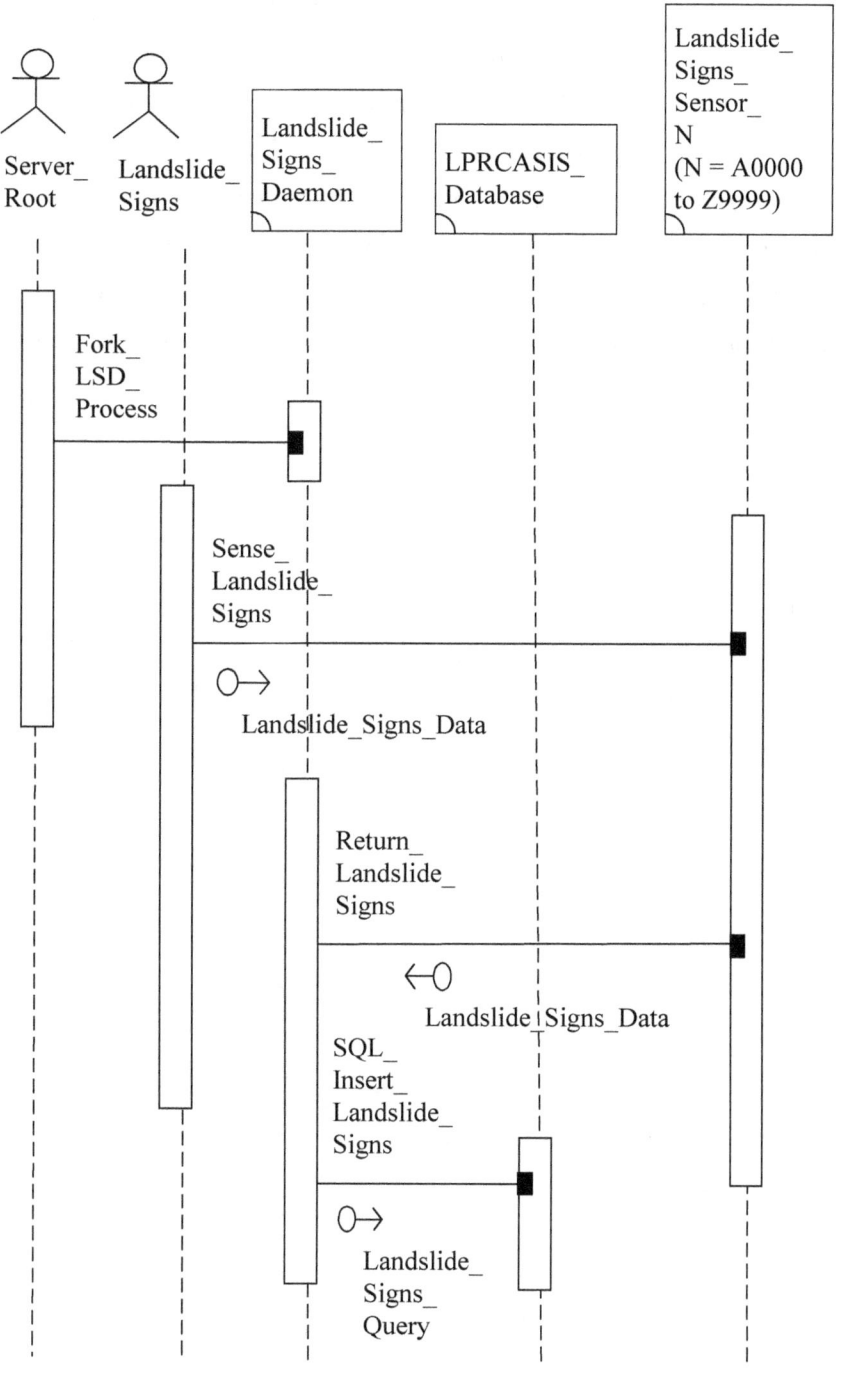

Figure 8-14   IFD of the *Sensing_Landslide_Signs* Behavior

Figure 8-15 shows an IFD of the *Recording_Occurring_Landslide* behavior. First, actor *Disaster_Supervisor* interacts with the *Occurring_Landslide_UI* component through the *Input_Occurring_Landslide* operation call interaction, carrying the *Occurring_Landslide_Form* input parameter. Last, component *Occurring_Landslide_UI* interacts with the *LPRCASIS_Database* component through the *SQL_Insert_Occurring_Landslide* operation call interaction, carrying the *Occurring_Landslide_Query* input parameter.

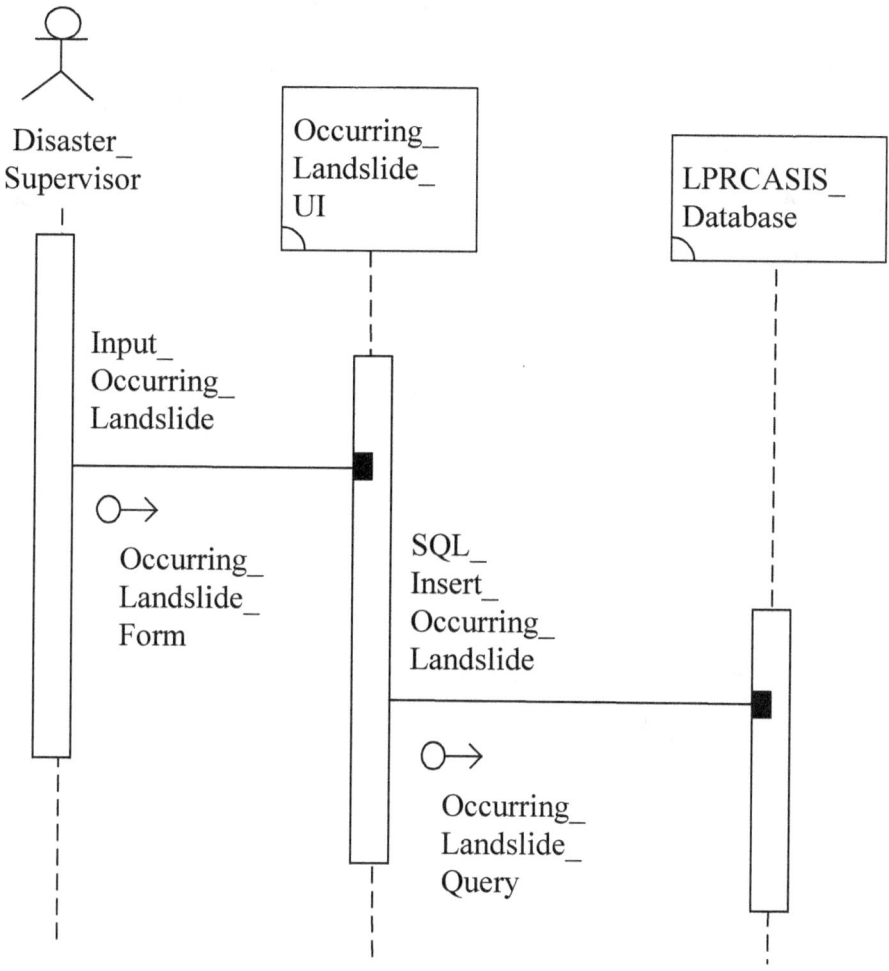

Figure 8-15   IFD of the *Recording_Occurring_Landslide* Behavior

Figure 8-16 shows an IFD of the *Alerts_Notifying* behavior. First, actor *Two_Minutes_Interval* interacts with the *Alerts_Notifying_UI* component through the *Showing_All_Alerts* operation call interaction, carrying the *Current_Time* input parameter. Next, component *Alerts_Notifying_UI* interacts with the *LPRCASIS_Database* component through the *SQL_Select_Landslides_Signs_for_Alerts_Analysis* operation call interaction, carrying the *Current_Time* input parameter and the *Landslides_Signs_for_Alerts_Analysis_Query* output parameter. Finally, actor *Disaster_Supervisor* interacts with the *Alerts_Notifying_UI* component through the *Display_Alerts* operation return interaction, carrying the *Alerts_AreaMap* output parameter.

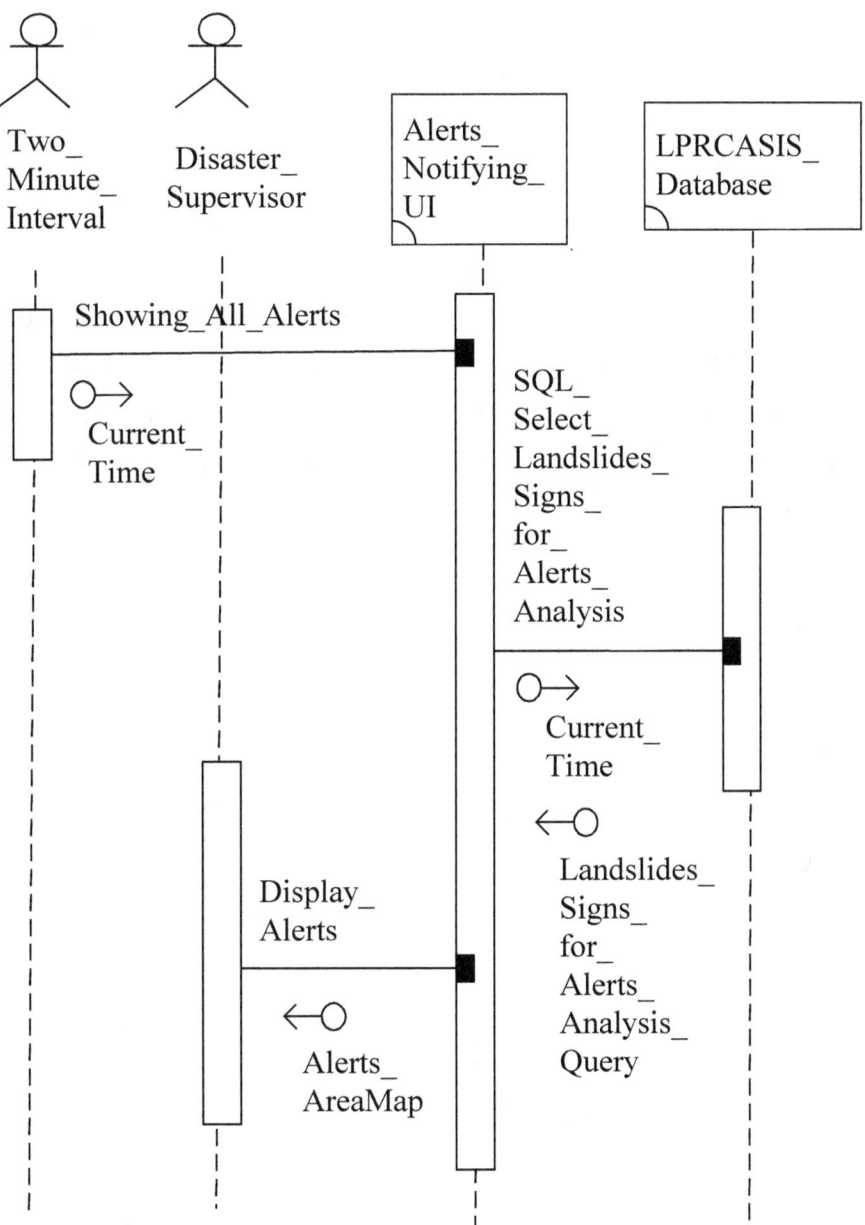

Figure 8-16   IFD of the *Alerts_Notifying* Behavior

Figure 8-17 shows an IFD of the *Recording_Emergency_Response_Starting_Time* behavior. First, actor *Disaster_Supervisor* interacts with the *Emergency_Response_Starting_Time_UI* component through the *Input_Emergency_Response_Starting_Time* operation call interaction, carrying the *Emergency_Response_Starting_Time_Form* input parameter. Last, component *Emergency_Response_Starting_Time_UI* interacts with the *LPRCASIS_Database* component through the *SQL_Insert_Emergency_Response_Starting_Time* operation call interaction, carrying the *Emergency_Response_Starting_Time_Query* input parameter.

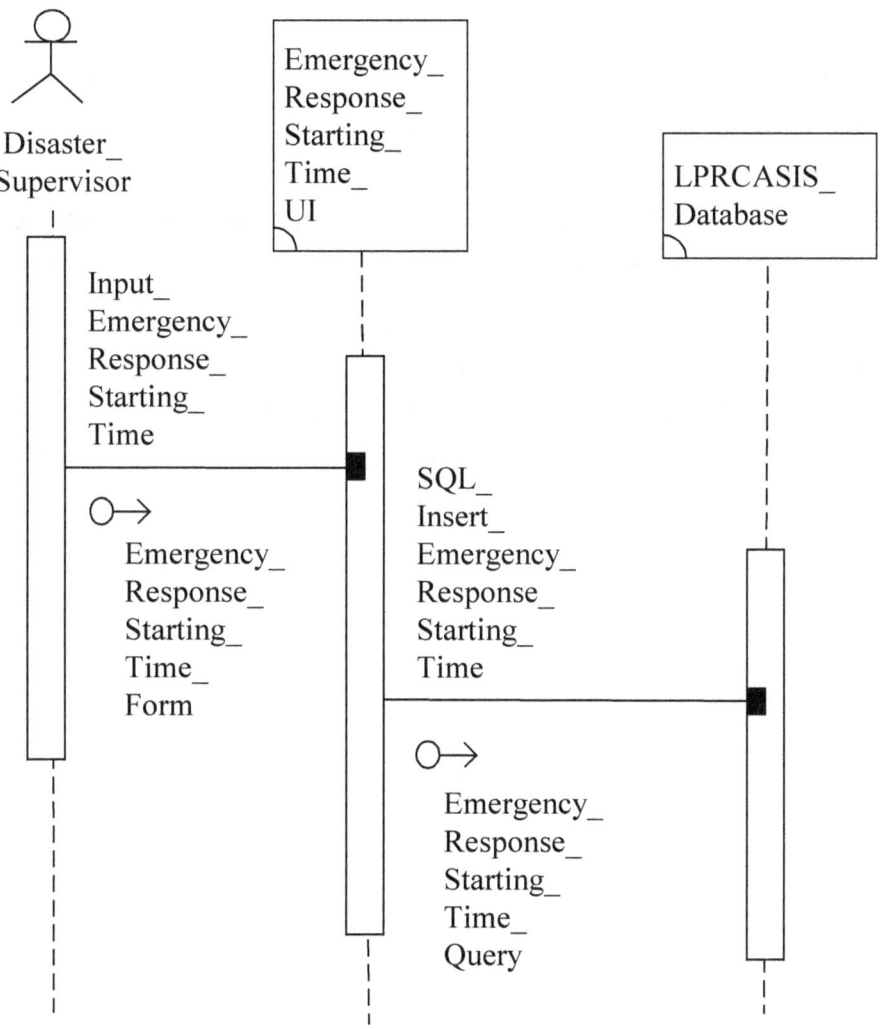

Figure 8-17   IFD of the
Recording_Emergency_Response_Starting_Time Behavior

Figure 8-18 shows an IFD of the *Recording_Emergency_Response_End_Time* behavior. First, actor *Disaster_Supervisor* interacts with the *Emergency_Response_End_Time_UI* component through the *Input_Emergency_Response_End_Time* operation call interaction, carrying the *Emergency_Response_End_Time_Form* input parameter. Last, component *Emergency_Response_End_Time_UI* interacts with the *LPRCASIS_Database* component through the *SQL_Insert_Emergency_Response_End_Time* operation call interaction, carrying the *Emergency_Response_End_Time_Query* input parameter.

137

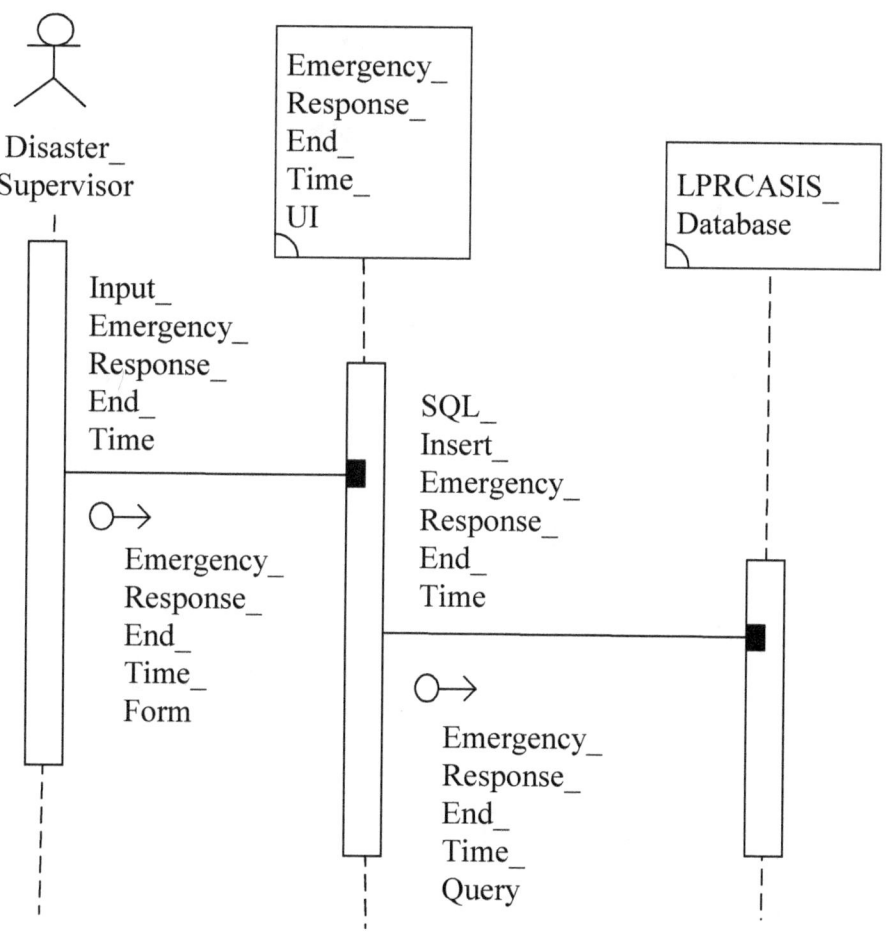

Figure 8-18   IFD of the
Recording_Emergency_Response_End_Time Behavior

APPENDIX A: SYSTEM REQUIREMENTS SPECIFICATION 2.0

(1) Architecture Hierarchy Diagram

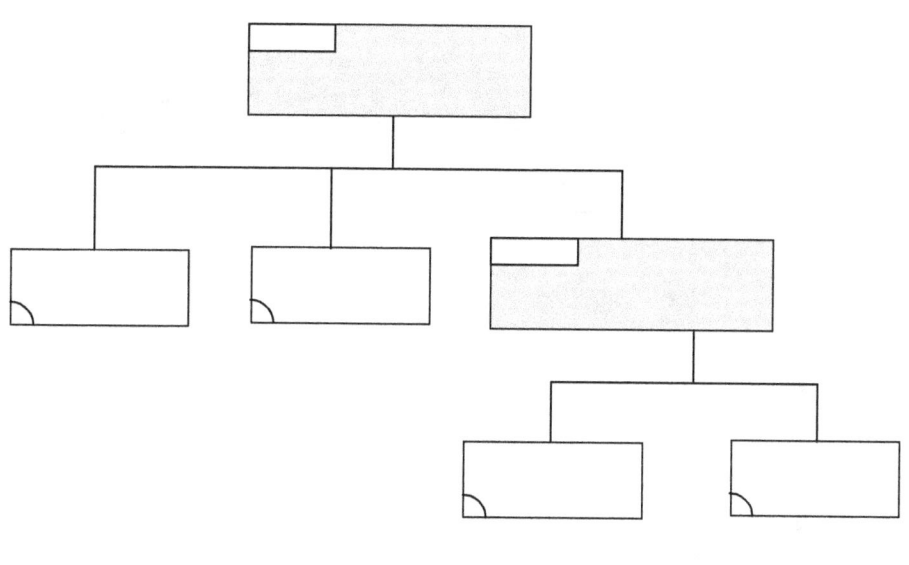

: Aggregated System

: Non-Aggregated System,   Component

(2) Component Operation Diagram

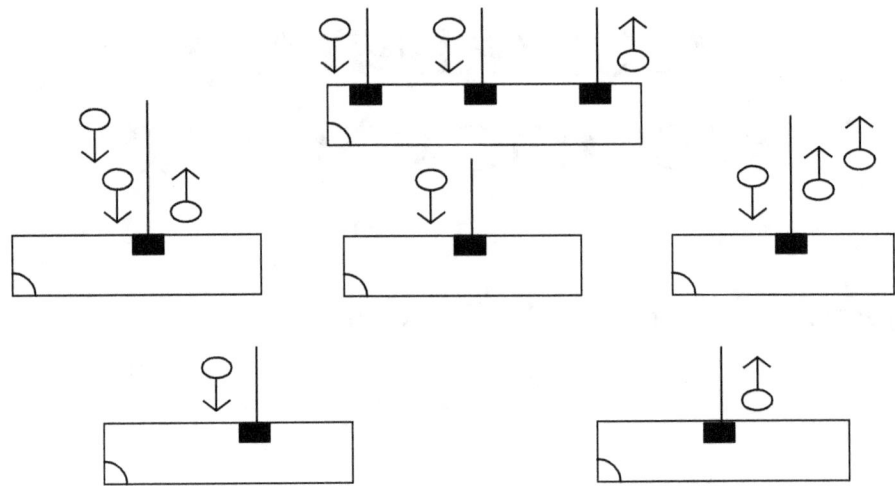

∎	: Operation
⊋↓	: Input Data
↑⊋	: Output Data
▭	: Component

(3) Interaction Flow Diagram

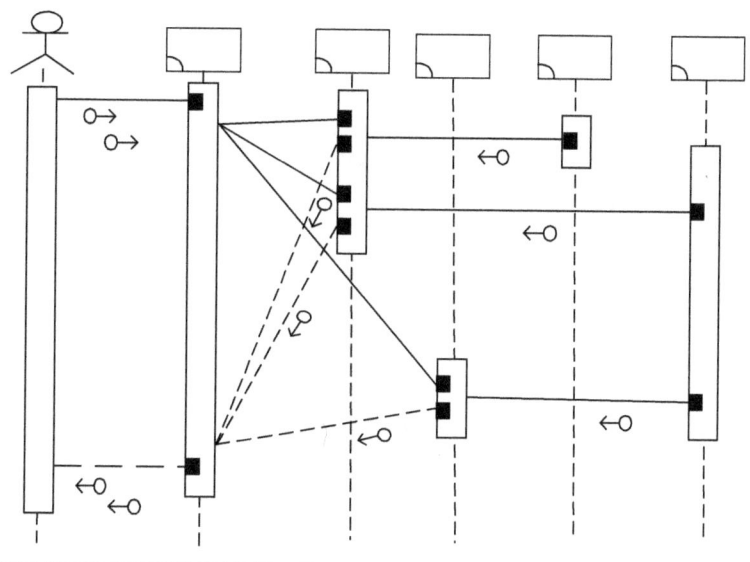

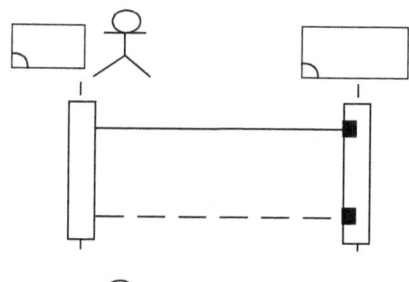

: Operation Call Interaction

: Operation Return Interaction

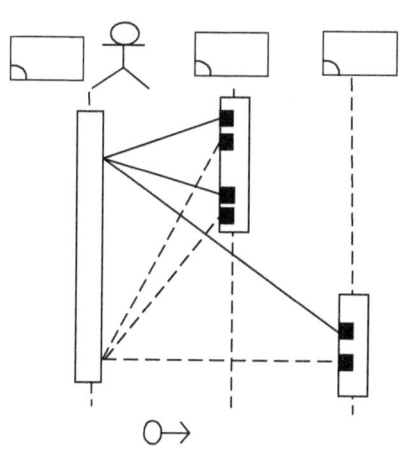

: Conditional
Operation Call Interaction

: Conditional
Operation Return Interaction

O→   : Input Data

←O   : Output Data

APPENDIX B: SBC PROCESS ALGEBRA

(1) Operation-Based Single-Queue SBC Process Algebra

(1) <System> ::= **fix(**" <Process_Variable> "="<IFD> " ● " <Process_Variable>
{"+" <IFD> " ● " <Process_Variable>} ")"

(2) <IFD> ::= <Type_1_Interaction> {"● " <Type_1_Or_2_Interaction>}

(3) <Type_1_Or_2_Interaction> ::= <Type_1_Interaction>

| <Type_2_Interaction>

(2) Operation-Based Multi-Queue SBC Process Algebra

(1) <System> ::= <FixIFD> {"‖ " <FixIFD>}

(2) <FixIFD> ::= "**fix**(" <Process_Variable>"="<IFD>
         "●" <Process_Variable> ")"

(3) <IFD> ::= <Type_1_Interaction> {"● " Type_1_Or_2_Interaction>}

(4) <Type_1_Or_2_Interaction> ::= <Type_1_Interaction>

         | <Type_2_Interaction>

(3) Operation-Based Infinite-Queue SBC Process Algebra

(1) \<System\> ::= "! ("\<IFD\> " ● " *STOP* ")" {"|| ! (" \<IFD\> " ● " *STOP* ")"}

(2) \<IFD\> ::= \<Type_1_Interaction\> {"● " \<Type_1_Or_2_Interaction\>}

(3) \<Type_1_Or_2_Interaction\> ::= \<Type_1_Interaction\>

   | \<Type_2_Interaction\>

BIBLIOGRAPHY

[Ashw90] Ashworth, C., *SSADM : A Practical Approach*, 1st Edition, McGraw-Hill Book Company (UK) Ltd., 1990.

[Bash86] Bashe, C., *IBM's Early Computers*, The MIT Press, 1986.

[Bern09] Bernstein, D. et al., "Blueprint for the Intercloud – Protocols and Formats for Cloud Computing Interoperability," *IEEE Computer Society*, 2009, pp.328-336.

[Bi06] Bi, Y. et al., "A Parking Management System Based on Wireless Sensor Network," *ACTA AUTOMATICA SINICA*, Vol. 32, No. 6, 2006, pp. 38-45.

[Blan08] Blanchard, B. S., *System Engineering Management*, 4th Edition, Wiley, 2008.

[Booc07] Booch, G., *Object-oriented Analysis and Design with Applications*, 3rd Edition, Addison-Wesley, 2007.

[Came89] Cameron, John R., *The Jackson Approach to Software Development*, IEEE Computer Society Press, 1989.

[Chao14a] Chao, W. S., *Systems Thingking 2.0: Architectural Thinking Using the SBC Architecture Description Language*, CreateSpace Independent Publishing Platform, 2014.

[Chao14b] Chao, W. S., *General Systems Theory 2.0: General Architectural Theory Using the SBC Architecture*, CreateSpace Independent Publishing Platform, 2014.

[Chao14c] Chao, W. S., *Systems Modeling and Architecting: Structure-Behavior Coalescence for Systems Architecture*, CreateSpace Independent Publishing Platform, 2014.

[Chao15a] Chao, W. S., *Theoretical Foundations of Structure-Behavior Coalescence*, CreateSpace Independent Publishing Platform, 2015.

[Chao15b] Chao, W. S., *Variants of Interaction Flow Diagrams*, CreateSpace Independent Publishing Platform, 2015.

[Chao15c] Chao, W. S., *A Process Algebra For Systems Architecture: The Structure-Behavior Coalescence Approach*, CreateSpace Independent Publishing Platform, 2015.

[Chao15d] Chao, W. S., *An Observation Congruence Model For Systems Architecture: The Structure-Behavior Coalescence Approach*, CreateSpace Independent Publishing Platform, 2015.

[Chao15e] Chao, W. S., *Variants of SBC Process Algebra: The Structure-Behavior Coalescence Approach*, CreateSpace Independent Publishing Platform, 2015.

[Chao17a] Chao, W. S., *Channel-Based Single-Queue SBC Process Algebra For Systems Definition: General Architectural Theory at Work*, CreateSpace Independent Publishing Platform, 2017.

[Chao17b] Chao, W. S., *Channel-Based Multi-Queue SBC Process Algebra For Systems Definition: General Architectural Theory at Work*, CreateSpace Independent Publishing Platform, 2017.

[Chao17c] Chao, W. S., *Channel-Based Infinite-Queue SBC Process*

Algebra For Systems Definition: General Architectural Theory at Work, CreateSpace Independent Publishing Platform, 2017.

[Chao17d] Chao, W. S., *Operation-Based Single-Queue SBC Process Algebra For Systems Definition: General Architectural Theory at Work*, CreateSpace Independent Publishing Platform, 2017.

[Chao17e] Chao, W. S., *Operation-Based Multi-Queue SBC Process Algebra For Systems Definition: Unification of Systems Structure and Systems Behavior*, CreateSpace Independent Publishing Platform, 2017.

[Chao17f] Chao, W. S., *Operation-Based Infinite-Queue SBC Process Algebra For Systems Definition: Unification of Systems Structure and Systems Behavior*, CreateSpace Independent Publishing Platform, 2017.

[Chen76] Chen, P. et al., "The Entity-Relationship Model - Toward a Unified View of Data", *ACM Transactions on Database Systems* 1 (1), pp. 9–36, 1976.

[Coro08] Corominas, J. et al., "A Review of Assessing Landslide Frequency for Hazard Zoning Purposes," *Elsevier Engineering Geology*, Volume 102, 2008, pp. 193-213.

[Date03] Date, C. J., *An Introduction to Database Systems*, 8th Edition, Addison Wesley, 2003.

[DeMa79] DeMarco, T., *Structured Analysis and System Specification*, Prentice Hall, 1979.

150

[Denn08] Dennis, A. et al., *Systems Analysis and Design*, 4th Edition, Wiley, 2008.

[Dori95] Dori, D., "Object-Process Analysis: Maintaining the Balance between System Structure and Behavior," *Journal of Logic and Computation* 5(2), pp.227-249, 1995.

[Dori02] Dori, D., *Object-Process Methodology: A Holistic Systems Paradigm*, Springer Verlag, New York, 2002.

[Dori16] Dori, D., *Model-Based Systems Engineering with OPM and SysML*, Springer Verlag, New York, 2016.

[Elma10] Elmasri, R., *Fundamentals of Database Systems*, 6th Edition, Addison Wesley, 2010.

[Fell08] Fell, R. et al., "Guidelines for Landslide Susceptibility, Hazard, and Risk Zoning for Land Use Planning," *Elsevier Engineering Geology*, Volume 102, 2008, pp. 85-98.

[Ghar11] Gharajedaghi, J., *Systems Thinking: Managing Chaos and Complexity: A Platform for Designing Business Architecture*, Morgan Kaufmann, 2011.

[Grad06] Grady, J. O., *System Requirements Analysis*, 1st Edition, Academic Press, 2006.

[Grad13] Grady, J. O., *System Requirements Analysis*, 2nd Edition, Elsevier, 2013.

[Hatl00] Hatley, D. J. 2t al., *Process for System Architecture and Requirements Engineering*, 1st Edition, 2000.

[Hoar85] Hoare, C. A. R., *Communicating Sequential Processes*,

Prentice-Hall, 1985.

[Hoff10] Hoffer, J. A., et al., *Modern Systems Analysis and Design*, Prentice Hall, 6th Edition, 2010.

[Kend10] Kendall, K. et al., *Systems Analysis and Design*, 8th Edition, Prentice Hall, 2010.

[Koss11] Kossiakoff, A. et al., Systems Engineering Principles and Practice, 2nd Edition, Wiley-Interscience, 2011.

[Lapl13] Laplante, P. A., *Requirements Engineering for Software and Systems*, 2nd Edition, Auerbach Publications, 2013.

[Marc88] Marca, D. A. et al., *SADT: Structured Analysis and Design Technique,* McGraw-Hill, 1988.

[Miln89] Milner, R., *Communication and Concurrency*, Prentice-Hall, 1989.

[Miln99] Milner, R., *Communicating and Mobile Systems: the π-Calculus*, 1st Edition, Cambridge University Press, 1999.

[Pele00] Peleg, M. et al., "The Model Multiplicity Problem: Experimenting with Real-Time Specification Methods". *IEEE Tran. on Software Engineering.* 26 (8), pp. 742–759, 2000.

[Pres09] Pressman, R. S., *Software Engineering: A Practitioner's Approach*, 7th Edition, McGraw-Hill, 2009.

[Reis92] Reisig, W., A Primer in Petri Net Design, Springer-Verlag, 1992.

[Scho10] Scholl, C., *Functional Decomposition with Applications to FPGA Synthesis*, Springer, 2010.

[Seth96] Sethi, R., *Programming Languages: Concepts and Constructs*, 2nd Edition, Addison-Wesley, 1996.

[Shel11] Shelly, G. B., et al., *Systems Analysis and Design*, 9th Edition, Course Technology, 2011.

[Sode03] Soderborg, N.R. et al., "OPM-based Definitions and Operational Templates," *Communications of the ACM* 46(10), pp. 67-72, 2003.

[Somm06] Sommerville, I., *Software Engineering*, 8th Edition, Addison-Wesley, 2006.

[Wald15] Walden, D. D. et al., *INCOSE Systems Engineering Handbook: A Guide for System Life Cycle Processes and Activities*, 4th Edition, Wiley, 2015.

INDEX

A

aggregated system, 63

AHD. *See* architecture hierarchy diagram

architecture hierarchy diagram, 57, 139

B

building block. *See* component

C

COD. *See* component operation diagram

component, 29

component operation diagram, 65, 70, 140

E

entity. *See* component

entity relationship modeling, 24

ERM. *See* entity relationship modeling

F

flow chart, 25

function. *See* operation

I

IFD. *See* interaction flow diagram

interaction, 20, 21

interaction flow diagram, 73, 141

internet of things, 83, 111

IoT. *See* internet of things

J

jackson system development, 24

JSD. *See* jackson system development

M

method. *See* operation

model multiplicity, 23

model multiplicity problem, 23

model singularity, 26, 51

multi-level, 59

    composition, 59

    decomposition, 59

multiple models. *See* model multiplicity

    behavior model, 23

    data model, 23

    function model, 23

structure model, 23

multiple views, 20

    behavior view, 20

    data view, 20

    function view, 20

    structure view, 20

multiple views integrated. *See* system
    requirements specification 2.0

multiple views non-integrated. *See* system
    requirements specification 1.0

N

non-aggregated system, 63, *See* component

O

object. *See* component

object-oriented analysis, 25

OOA. *See* object-oriented analysis

operation, 30, 65

P

part. *See* component

petri net, 25

problem space, 17

procedure. *See* operation

S

SA. *See* structured analysis

SADT. *See* structured analysis and design
    technique

SBC. *See* structure-behavior coalescence

SDLC. *See* systems development life cycle

single model. *See* model singularity

solution space, 17

SSADM. *See* structured systems analysis
    and design method

structure-behavior coalescence, 43, 46, 51

structured analysis, 25

structured analysis and design technique, 25

structured systems analysis and design
    method, 25

SyRS. *See* system requirements
    specification

system requirements specification, 15, 18

system requirements specification 1.0, 23

    control-oriented, 24, 25

    data-oriented, 24

    function-oriented, 24, 25

    object-oriented, 24, 25

system requirements specification 2.0, 27,
    50

    architecture hierarchy diagram, 57, 84,
        112, 139

    component operation diagram, 65, 86,
        114, 140

interaction flow diagram, 73, 101, 127, 141

systems analyst, 18

systems behavior, 20, 35

systems development life cycle, 15

   design and implementation, 17

   product evolution, 17

   project planning, 16

   requirements and specifications, 17

   verification and validation, 17

systems structure, 20, 29

V

V&V. *See* verification and validation

verification and validation, 17

www.ingramcontent.com/pod-product-compliance
Lightning Source LLC
Chambersburg PA
CBHW081452170526
45166CB00008B/2404